MW01470744

# The *New* Manufacturing Engineer

# The *New* Manufacturing Engineer

## Coming of age in an agile environment

## by Michael J. Termini

Published by
Society of Manufacturing Engineers
One SME Drive, P.O. Box 930
Dearborn, Michigan 48121-0930

Library of Congress Catalog Card Number: 96-069144
International Standard Book Number: 0-87263-479-5


Additional copies may be obtained by contacting:

Society of Manufacturing Engineers
Customer Service
One SME Drive
Dearborn, Michigan 48121
1-800-733-4763

SME staff who participated in producing this book:

Donald A. Peterson, Senior Editor
Dorothy M. Wylo, Production Assistant
Frances M. Kania, Production Assistant
Rosemary K. Csizmadia, Production Administrator
Karen M. Wilhelm, Manager, Book Publishing
Cover design by Judy D. Munro, Manager, Graphic Services


Printed in the United States of America

This book could not have been written without the love and support of my wife, Susan, and the patience of my children, Kelly, Justin, Casey, and Brad. My love and heartfelt thanks to each of you.

My thanks, too, to my parents, Joe and Wanda, for teaching me the value of hard work, and for giving me the courage to pursue new avenues, dreams, and challenges. Dad, your memory lives on in me each day. You will never be forgotten.

# Table of Contents

# Preface

Traditional manufacturing is rapidly giving way to new, fast-response, customer-focused techniques that maximize the manufacturer's return on all resources—capital, materials, equipment, facilities, personnel, and most importantly, *time*. Farsighted manufacturers today are implementing cultural changes throughout their organizations that complement technological advancements and facilitate the transition from mass, to lean, to agile manufacturing techniques. Individual contribution is giving way to self-managed and self-directed employee teams utilizing empowerment to manage day-to-day floor activities and to resolve production issues.

Concurrent engineering techniques are replacing the classical design engineering cycle, yielding 40-50% reductions in new product development cycle times, with corresponding reductions of 35-45% in resources employed.

Core process and organizational re-engineering are effectively restructuring the way products are designed and produced, as well as the processes and organizations that support them.

Customer expectations of quality are growing higher each day, while tolerance of perceived excessive costs and lead times is rapidly diminishing. The customer's buying decision is moving away from purchase price to total life-cycle costs of ownership, driving manufacturers to consider maintainability, upgradability, and recyclability in their product planning and development. Change is prevalent in every facet of the manufacturing sector, and any organization not involved in the change process has made a strategic decision to be out of business within the next 8 to 10 years.

This incredible change requires a new breed of manufacturing engineer. No longer merely the manufacturing technician, the manufacturing engineer of tomorrow will be more expansive, more people-oriented, more facilitation focused. The new manufacturing engineer must become a total enterprise strategist, tactician, and technician, a leader who:

- Fully understands new product design and development;
- Advises management on strategic product and operational issues and their associated risks;

- Designs material handling, processing, and storage systems utilizing the latest agile manufacturing techniques;
- Specifies and procures capital equipment and outside processing consistent with the organization's current and projected core competencies;
- Acts as a team player—and leader—in the selection and certification of strategic suppliers in the supply-chain management processes;
- Manages hazardous waste distribution and disposal for compliance with all new and emerging federal, state, and local regulations;
- Designs and administers workplace safety processes and conducts relevant employee training to ensure compliance with the processes;
- Advises management on potential product liability issues;
- Provides financial and performance data for monitoring quality costs, product costs, and operational performance metrics;
- Understands global business dynamics, emerging technologies, sourcing, and competitive issues;
- Serves as mentor, facilitator, and educator to the workforce;
- Advises on product and process capabilities and limitations to ensure that products are designed in harmony with readily available internal and external capabilities;
- Understands and advises on logistical considerations; and
- Oversees the manufacturing planning, confirmation, and operational processes in context with corporate strategic plans.

My intent in writing this book is to provide today's manufacturing engineers with the fundamental principles and practices to guide their organizations into the millennium . . . to compete successfully . . . to design and build products that are environmentally benign and customer-specific . . . and to maximize return on investment and assets. In short, to thrive in tomorrow's dynamic, global marketplace.

Michael J. Termini
West Palm Beach, Florida 1996

# 1

# An Expanded Role with New Responsibilities

## THE MANUFACTURING ENGINEER AS STRATEGIC PLANNER

As industry evolves from traditional mass production into flexible/lean manufacturing, then into agile manufacturing techniques, so must the manufacturing engineer evolve into a multidisciplined strategic planner of both business and operational tactical goals and objectives. A slow, orderly, evolutionary transformation, however, is not feasible because of the ever-changing dynamics of both global competition and the speed of technological advancements. The requirement for a rapid transformation of the manufacturing engineer's role is being fueled by:

- Increasing product and process sophistication and variability;
- Rapid and continuous advancements in technology;
- A changing focus toward total life-cycle costs by the consumer versus a singular purchase price;
- Global competition that is becoming increasingly *time* conscious; and
- A multitude of environmental, social, and economic pressures driven by regulatory dictates within both the domestic and international venues.

## CHANGING ENVIRONMENTS, CHANGING ROLES

Change is a natural process. But for the traditional manufacturing engineer, dealing with change can be particularly traumatic. No longer just the operations and manufacturing technician, the manufacturing engineer of the 21st century will be forced into a role demanding a far broader complement of skill sets. Skills that include the ability to function as:

- *A manufacturing operations strategist*, capable of blending both the corporation and its supply base into a stable infrastructure which supports the *strategic direction* of the organization as it competes in increasingly dynamic markets.
- *A business strategist*, with the capacity to quickly and effectively translate business and marketing plans into specific product, process, and facility requirements to ensure that the organization's business and market objectives are achieved within the time frames and budgetary constraints required to guarantee success.
- *A technology visionary*, capable of accurately forecasting technological trends and implementing state-of-the-art methodologies that will keep the organization competitive in operating cost, product quality, inventory investment, and cycle time.
- *A project manager* with the ability to comprehend a host of multifunctional business problems, and manage the activities of multidisciplinary project teams to ensure the most time- and cost-effective problem resolutions.
- *A team leader*, as well as a team player, with the capacity and willingness to participate in an empowered environment where decision making is driven as deep into the organizational structure as possible. No longer the director, the new manufacturing engineer will be a people-oriented facilitator capable of listening to diverse opinions and formulating corrective actions that incorporate those views into the ultimate solution.
- *An educator and trainer*, a mentor and counselor: the "informal leader" who shares expertise and knowledge with others for the betterment of the entire organization.

Changes in the manufacturing environment will also dictate the necessity for the manufacturing engineer to assume responsibilities previously within the exclusive domain of other functions, such as:

- Supplier selection, development, and certification;
- Product and process design, using concurrent engineering techniques to reduce the new product development cycle-time and associated development costs;
- Assessment of projected total product life-cycle costs and the contribution of the associated manufacturing processes to those costs;

- Organizational and core process re-engineering to reduce operational and administrative waste, reduce total cycle time, and minimize product and nonproduct costs;
- Identifying and consummating strategic alliances with critical suppliers, technology developers, customers, direct and indirect competitors, and market leaders;
- Isolating market and technological niches where future growth and success can be assured;
- Guiding total quality management initiatives—in production, design, procurement, customer service, administration—in short, all areas within the organization that impact the manufacturing operations;
- Implementing and participating in self-directed work teams and other employee-involvement processes;
- Using sound design-for-manufacturability, -assembly, and -maintainability techniques to ensure that all products can be manufactured within the targeted price-point range with readily available process capabilities;
- Implementing preventive, predictive, and productive maintenance methodologies to ensure that all process and manufacturing equipment, tooling, and fixtures are maintained in conformance to the manufacturers' specifications. By so doing, the capability of producing quality products at rated operating speeds is ensured, and the shortest possible changeover cycle times are guaranteed; and
- Introducing technological advancements in both products and processes as dictated by unanticipated opportunities and market and competitive pressures.

## TECHNOLOGICAL GROWTH—THE CHANGE AGENT

A study conducted by the Society of Manufacturing Engineers, A.T. Kearney, and Lehigh University illustrates the projected growth in those technologies that will redefine both the role of the manufacturing engineer and the strategic direction of the organizations they serve in the coming years.[1] These changes in technology will become the drivers of individual growth and development of tomorrow's manufacturing engineers. As you read through the projections in Table 1-1, consider how far you and your organization have progressed technologically, and how much more needs to be done to achieve and maintain world-class status.

Couple the growth projected in Table 1-1 with that of the quality and operational advancements recorded during the last decade (Table 1-2), and the challenge becomes abundantly clear.

The critical question is, are you part of this change movement, or are you still relying on what you did in the past to get you through? Given the rapid changes in both market and customer dynamics, those methods may not even guarantee your organization parity with your past successes. The business risks associated with following the old paradigms are significant. Lead, follow, or fail, the choice is yours.

### Table 1-1. Projected Growth
### in Manufacturing Technologies—1992 to 2000

| | |
|---|---|
| Expert systems | 36% |
| Artificial intelligence | 36% |
| Networking | 36% |
| Automated material handling | 35% |
| Sensor technology | 35% |
| Machine vision | 35% |
| Laser applications | 34% |
| Integrated manufacturing systems | 30% |
| Flexible manufacturing systems | 24% |
| Simulation and modeling techniques | 23% |
| Composite materials | 20% |
| CAD/CAM/CAE/EDI/Bar Coding | 13% |
| Manufacturing in space | 11% |

### Table 1-2. Growth in Quality Initiatives—1985 to 1995

| | |
|---|---|
| Statistical process control | 86% |
| Employee empowerment and teams | 77% |
| Concurrent engineering | 68% |
| Supplier partnerships and alliances | 67% |
| Total cycle time management | 66% |
| Core process re-engineering | 58% |
| Business-based training of employees | 54% |

# WHAT, THEN, IS TOMORROW'S MANUFACTURING ENGINEER?

The 21st-century manufacturing engineer will be a hybrid, a business strategist and an operations tactician who:

- Understands and actively participates in product design and development through the concurrent engineering process;
- Advises management on strategic issues, along with their associated technical and process risks;
- Designs material processing, assembly, and handling systems to maximize production throughput, while minimizing change-over times;
- Specifies and procures capital equipment to ensure technological advantage and competitiveness through process capability and control;
- Develops effective setup reduction strategies and preventive and predictive maintenance processes to maximize equipment uptime;
- Manages hazardous waste distribution and disposal to ensure compliance with Environmental Protection Agency (EPA) and other regulatory requirements;
- Administers workplace safety processes as defined by Occupational Safety and Health Administration (OSHA) and industry guidelines;
- Advises on product liability and life-cycle issues;
- Provides financial and performance data for monitoring operational, quality, and product cost metrics;
- Studies and analyzes global business and competitive issues to ensure the correct operational and strategic focus is maintained;
- Acts as mentor, facilitator, and educator of the work force to foster cross-training, cross-functional creativity, and teamwork;
- Advises on product and process capabilities and limitations to ensure the proper alignment between product design and process capabilities;
- Advises operations management on logistical considerations relative to their influence on cycle time, transportation damage, and product costs; and
- Oversees the production processes and the outside influences that can affect them.

To accomplish this extraordinary role change, manufacturing engineers will no longer work alone. There will be no manufacturing or engineering superstars in the new manufacturing arena, only super teams supporting winning organizations . . . super teams possessing complementary skill sets, common goals and metrics, a universally embraced and supported approach, and a synergistic chemistry that promotes focus on organizational successes as opposed to individual achievement.

## Reference

1. A.T. Kearney, Society of Manufacturing Engineers, and Lehigh University; *Profile 21*, 1988. Dearborn, Mich., SME.

# 2

# Reacting to Dynamic Customer Requirements

It has been said that quality is secondary if your products are not introduced to the market at the proper time and at a perceived price/value ratio that meets the customers' expectations. Market opportunity is fleeting. Unless your organization acts expeditiously, the opportunity will be seized by the competition.

## THE IMPORTANCE OF TIMING

Remember the crock pot, that interesting device into which you throw fresh vegetables and meats in the morning and let it cook all day while you are at work? When you return home, dinner is ready—no muss, no fuss. A great idea for today's busy family. Do you remember who initially advertised the concept of the crock pot? It was Rival Manufacturing. But before Rival could effectively exploit its new product (due to manufacturing delays), the market was filled with crock pots from numerous manufacturers. As a result, Rival lost a tremendous opportunity to capture a new and lucrative market, despite its being the product's creator.

The scenario is all too common in industry. Great ideas turn into lost opportunities if the products cannot be brought to market in time. Global competition is ruthless. Winning requires more than just great ideas, it requires great execution. Thus, time is the critical element of the new-product equation—today, and certainly in the millennium. The world-class manufacturing engineer must therefore focus his or her talents and attention on maximizing the organization's return on invested time by eliminating all waste and nonvalue-added activities within the enterprise. With that in mind, respond to the questions in Table 2-1.

### Table 2-1. Optimizing Time to Market

| |
|---|
| *Question 1.* Are your bills of material, product specifications, process routings, and product costs 100% accurate? <br> ☐ Yes       ☐ No <br><br> *Question 2.* What percentage of each workday is spent on truly value-added activities, as defined by your job description? <br> ☐ 100    ☐ 90    ☐ 80    ☐ 70    ☐ Less |

Look at each of your answers to determine how your current operations are impacting your organization's total value-added time.

Question 1. If your bills of material, product specifications, routings, etc. are not 100% accurate (and most aren't), how can you expect your employees and suppliers to consistently produce 100% conforming materials, in the shortest amount of time, and at the lowest total cost to your organization? How can you expect your systems to provide you with accurate, timely data on which to base critical business and strategic decisions, recognizing that these are the bases from which all system explosions are generated? These systems are the foundation on which all calculations within your inventory planning, process routing, and product cost systems are based. Put simply, how can you expect your direct and indirect support systems to operate if you can't even tell them exactly what it is you want?

Question 2. Everyone is overworked. Few, if any, companies today have excess resources standing idly by waiting for something to do. No one has time to do things right the first time, so they are done quickly and often haphazardly. In most organizations, manufacturing engineers (and most other disciplines) spend between 75-85% of every day doing something other than what's in their job descriptions . . . fire fighting, rework processing, attending material review board meetings, waiting on information or correcting the information when it is finally received, expediting materials or equipment from suppliers who have failed to deliver as promised, performing value analysis to reduce the costs on existing product or process designs, and so on.

What if every hour of every day was 100% productive for every manufacturing engineer? For every employee, for that matter? Not feasible? Why not? Ever marvel at how things get done quickly and effectively when "the barriers" are removed? In emergencies, things happen. Every department pulls its own weight. Approvals are circumvented, leaving the people in the trenches to make the appropriate decisions and implement the appropriate corrective actions. Suppliers become involved up front, and thus seem to perform as part of the team. Design, process, manufacturing, and quality engineers work together toward the desired outcome instead of against each other for individual political gains. Operators, supervisors, maintenance personnel, inspectors, buyers, planners—everyone—contributes, so the outcome is "owned" by everyone. No selling or politicking is required. Wow, what a great organization to work in! Then, it's business as usual. We forget those things that made the effort amazingly successful, and go back to the old paradigms and old results. Why not do it the right way every time?

Successful companies do just that. At Motorola, for instance, management has established key initiatives that focus on improving quality in all areas by defining precise customer-supplier performance metrics, then implementing monitoring techniques to ensure the desired performance is achieved, maintained, and improved upon. It is the foundation of their widely-recognized *"Six Sigma"* * total quality program. In addition, they concentrate their efforts in every department on reducing the cycle time of each activity. This is done by re-engineering their core processes to eliminate all processes and activities that do not add value to Motorola's customers, internal and external. In short, Motorola's focus is on maximizing total quality and minimizing cycle time by effectively eliminating those elements of their business that introduce waste into their *direct* and *indirect* processes. By so doing, they are able to maintain high quality processes and low operating costs, giving Motorola an enviable competitive edge.

Do your answers to the two questions indicate that there may be opportunities to improve your current operations? If so, let's explore in more detail some of the advanced techniques and methodologies used by many of the recognized world-class organizations as they prepare for the global competition of the 21st century.

*3.4 defects per million parts.

## MASTERING WORLD-CLASS PRACTICES

Any discussion of 21st-century manufacturing must begin with an explanation of the world-class manufacturing concepts that define the new industrial paradigm.

### Agile Manufacturing

Agility is not just another new word cloaking old practices, nor is it another of management's "flavor of the month" programs. Agility goes several steps beyond today's techniques of flexible and lean manufacturing by focusing attention on the core competencies a manufacturing enterprise must possess to be competitive in the global markets of the 21st century. Those competencies include:

- *Timely and consistent technological innovation* to support the organization's strategic direction and objectives . . . a direction that is constantly being redefined by the dynamics of the markets served;
- *The ability to develop and maintain a broad-based, well-educated work force* that is trained, involved, and ultimately empowered to make day-to-day decisions at all levels within the organization;
- *Enhanced communications and data processing networks* capable of linking all operating entities on a real-time basis, while providing direct access to external information and data sources through available domestic and international "information highways" like the Internet;
- *The ability to quickly identify potential market opportunities*, then to form strategic alliances with key suppliers and technology developers, customers, direct and indirect competitors, and/or manufacturers outside your industry segments to address those market opportunities in a cost- and time-sensitive manner. These alliances capitalize on the synergistic strengths of all partners to effectively outperform and outgun competition;
- *The ability to provide low-cost customized products* quickly and cost-competitively, through the use of modularized production facilities, tooling, equipment, and bills of materials, coupled with sound concurrent engineering techniques that focus on minimizing the new product development cycle time; and

- *The ability to provide products that are "life-cycle driven,"* based on sound design rules that address issues of producibility, assembly, confirmation, maintainability, upgradability, disassembly, and recyclability.

The differences between today's manufacturing methodologies and those within an agile environment (Table 2-2) are significant. The challenge of the new manufacturing engineer is to prepare his or her organization for the required changes in those core competencies *now*, and not wait until the competition, the customer, or the market forces you to react. By then, it may well be too late.

### Table 2-2. From Traditional Manufacturing to Agile

---

*1960s-1970s Traditional Manufacturing*

Incorporation of TQM and Quality Circles highlighted this attempt to initiate and integrate the desired characteristics of the Japanese manufacturers into the U.S. and Western European manufacturing communities.

*1980s Focused Factories*

Using just-in-time (JIT) theories, many U.S. manufacturers focused production into a few product families in an attempt to capture the advantages of group technologies, standardization of products and processes, and common tooling.

*1990s Flexible and Lean Manufacturing*

America's first dedicated involvement in supply-base management and empowerment techniques, this approach was aimed at capitalizing on the strengths inherent in the suppliers and employees of the company. It was the first comprehensive look at the entire value chain.

*2000s Agile Manufacturing*

The incorporation of supply-base approaches of concurrency, re-engineering, and total cycle-time management, with a focus on *time* as the strategic weapon. To minimize the rising costs of technology, agility incorporates the formation of strategic alliances to gain the advantages of rapid technological advancement without its associated investment.

---

As illustrated in Table 2-2, the transition from traditional manufacturing to agile manufacturing has steadily progressed over the last 30 years. It's an evolutionary move, not a revolutionary one.

Incorporation of world-class techniques like total quality management, JIT, supply-base management, supplier certification, employee empowerment, self-directed work teams, concurrent engineering, core process re-engineering, strategic alliances, and total cycle-time management have focused the manufacturing engineer's attention more pointedly on every aspect of the business that manifests itself into, or contributes to wasted time.

For the past several years, the emphasis in the manufacturing sector has been on flexible manufacturing techniques. The transition to agile manufacturing builds on those same principles while expanding on those elements that contribute to the reduction in both operational and administrative cycle times. Let's look at some of the major differences.

With flexible manufacturing, the goal is to minimize work-in-process (WIP) inventories by keeping as much inventory as possible in the raw state, thereby giving the organization the flexibility to convert raw materials into any number of final configurations to correspond with the customer orders received. The manufacturing engineer's concentration is on reducing in-process throughput cycle times to meet near-term customer requirements. Agile manufacturing takes that concept one step further by focusing the manufacturing engineer's attention on the reduction of all inventory categories—raw, WIP, and finished. Agility alters the manufacturing envelope further by defining products as "virtual commodities." In other words, there is no product definition until the customer defines a specific requirement by placing an order. While this may seem unrealistic, it is nonetheless plausible and workable. The case study on the facing page illustrates how a valve manufacturer has effectively applied these agile manufacturing techniques. Please read it before continuing.

If you were in the valve business, how would you compete?

Flexible manufacturing methodologies stress the elimination of waste throughout the manufacturing processes. Agile manufacturing, conversely, focuses the manufacturing engineer's attention on the elimination of waste in all areas of the business, thereby reducing the total cycle time rather than just the manufacturing cycle time. In the process, overhead is lowered as administrative and indirect labor requirements are reduced because the related tasks are performed more efficiently with fewer personnel.

### The Story of Ross Operating Valve Company

Over a 7-year period, Ross Operating Valve Company developed a true computer-integrated manufacturing (CIM) system focusing on quick changeover technologies. The $30-million investment in automated production equipment and CAD systems appears to have paid off, giving Ross the ability to produce customized valves practically overnight.

Ross boasts that it sells "virtual products": products that do not exist until the customer places the order. Ross's 15-person concurrent engineering team, working with the customer via computer link, designs sophisticated pneumatic valve assemblies to serve specialty applications. During the design process, all of the tool paths for machining and manufacturing the valve on advanced CNC machines are prepared. Upon completion, the data is downloaded to a waiting machine for production and subsequent assembly.

Paper drawings have been eliminated, as have manufacturing engineers, machinists, and inspectors. By eliminating many of the time consumers in the process, Ross can produce working prototypes of a customized valve within 72 hours. The customer is allowed to make a continuous stream of changes to the product at very little cost. And best of all, because the customer is part of the design team, he or she takes instant ownership of the design. As Ross puts it "we're no longer just a supplier; we're a partner" (a *strategic* alliance).

Ross also uses satellite links to pass information to its plants in Germany, England, Japan, and Troy and Madison Heights, Michigan.

In the future Ross plans to give its customers direct access to its Intergraph CAD terminals via phone so that they can design their own valves on Ross's terminals. Ross will provide the pneumatic technology; the customer will provide the application data. The valve will be shipped the next day.

Flexible manufacturing is based on the concept of flexibility in altering the master production schedule to accommodate minor near-term shifts in customer demand. Agile manufacturing forces the manufacturing engineer to structure both the direct and indirect processes in such a way as to provide the ability to build today what was sold yesterday—a macro *kanban* approach at the finished-goods level. Under such a premise, the inherent problems with inaccurate forecasting are eliminated.

With flexible manufacturing, lead times are based on industry standards, while with agile manufacturing, lead times are constantly reduced until they meet or exceed the customers' expectations. This constant reduction is a derivative of the organization's efforts to reduce cycle time in all direct and indirect areas. To illustrate, calculate the time it takes to process a customer's order through your sales and order-entry processes, then through the planning and procurement cycles for final release to Manufacturing. Then, compare that combined cycle time to the time it takes to actually produce your products. In many cases (most, in fact), it takes longer to process the customer's order than it takes to produce the product. Total cycle-time management forces the manufacturing engineer to minimize *all* elements of lead time in order to provide the customers with the shortest possible delivery time. This minimizes their (and your) exposure to post-release changes, while reducing the customer's inventory investment over the entire value chain.

Today, quality is often predefined as a subset of some nebulous industry standard, levels set by the *producers*, not the customers. With agile manufacturing, however, the baseline for quality is defined by *customer expectations*, both implicit and explicit.

Quality is also based on the total life-cycle costs of ownership, or use of your products in the customers' applications. It is, in essence, 100% conformance to the customers' stated and expected requirements; not 98%, not 99%, but 100%. Think of it this way: Would you accept anything less than 100% when you step on the brakes of your automobile? When you take an airline flight? When you agree to open-heart surgery? I think not.

That conformance includes product quality, reliability, performance, upgradability, on-time delivery, accuracy of quantities shipped against that ordered, and total cost. And quality must be consistent, as well as economically feasible. How many times has a supplier said to you, "You can have any quality level you want, so long as you are willing to pay for it."? Yet, the quality gurus tell us that quality is free. How can this be? Simply put, if a supplier has processes that are under total process control, and if those processes are statistically capable of producing products to your specifications, then they will make no bad products. The supplier's costs will be low because there will be no scrap, rework, or rejection costs, and those savings can be passed along to you, the customer. Therefore, quality can be free. We will explore this more later.

With flexible manufacturing, unit costs are kept low by producing large volumes of the same or similar products. With agile manufacturing, conversely, low unit costs are derived from modular production documentation, equipment, and facilities that can be configured as needed to economically meet unpredictable market demands, supported by rapid setup and changeover techniques. To enable this all to function smoothly, agile manufacturing requires an integrated enterprise-wide data and information system to quickly process and transmit needed data to the manufacturing and support users on a real-time basis.

## Concurrency

The concept of concurrency will be explained in more detail in later chapters. For now, it is important to recognize that concurrency is fundamental to reducing the time to market of new products and to the strategically planned introduction of enhancements to existing offerings. By definition, concurrency is the creation of a culture in which all functional disciplines that contribute to the design, development, production, distribution, and sale of a product perform their functions simultaneously versus sequentially, thereby reducing the cycle time to its lowest possible level by designing the product and its associated production processes at the same time.

Concurrency focuses the organization's attention on both the design and process requirements during the early concept development and preliminary design stages, where the result of a change is least costly and the impact on the targeted market introduction times is minimized. Economic and production feasibility is determined up front where numerous options are available to management, versus just prior to the published introduction date when there is little choice but to introduce less than optimum products at much less than anticipated profit margins. Quality is designed into the product and the process concurrently, thus guaranteeing that the product is both producible and economically feasible, while conforming to manufacturing, test, and customer requirements *before* the design is released to Manufacturing.

## Strategic Alliances

Today, more than ever before, we hear about the importance of aligning our organizations with capable, reliable suppliers and technology developers, by forming long-term relationships that are mutually beneficial to all parties. The old adversarial approach in dealing with suppliers, customers, and our own workforce is no longer appropriate, and typically leads to increases in cycle times and quality costs. Why? Because people are not working together with a common purpose and a common objective, nor with the belief that the outcome will be a "win-win" for all parties. Today, unless the objective is truly to benefit all parties, it just isn't going to be successful.

By definition, a strategic alliance is an association of producers, consumers, suppliers, and technology developers established to further the common interests of the individual members through an approach and methodology that addresses a market opportunity by identifying, attacking, and neutralizing the economic, operational, performance, political, and market strengths of the competition. One of the most common types of strategic alliance today is derived through supplier partnering by employing an effective supply-chain management process. In Chapter 8 of this book, we discuss the tools available to the manufacturing engineer to identify, develop, and nurture a strategic supplier alliance for the benefit of both the customer and the supplier.

### Alignment

The role of the manufacturing engineer is to assess a potential alliance partner's process capabilities as they relate to the needs and requirements of his or her own organization. Remembering that those needs are often dynamic, it is essential that the manufacturing engineer select alliance partners whose process and technological capabilities parallel the strategic plans established by your senior management relative to product and market expansion and penetration objectives.

The concept of alignment provides a straightforward look at the matching of customer requirements to supplier capabilities. When a supplier's capabilities, as measured by the tools developed in subsequent sections of this book, fall short of the customer's requirements, a "supply gap" results (Figure 2-1). This means that no matter how diligently a supplier works to meet its customer's requirements, the supplier will fall short of the customer's expectations. Not necessarily because of a lack of desire or dedication on that supplier's part, but because the supplier's processes are simply not statistically capable of meeting the requirements established for the customer's products.

This does not mean that the supplier is not qualified under other circumstances; it is just that for this particular product or service,

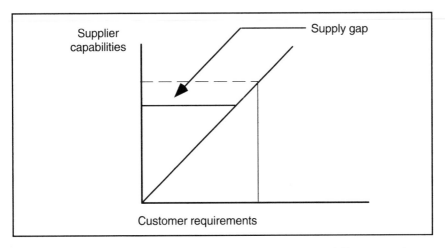

*Figure 2-1. When customer demands exceed supplier capability, a supply gap results.*

his or her processes are incapable of meeting your needs. The result of dealing with an incapable supplier will be ongoing nonconformances, increased quality costs, and increased cycle times resulting in delays in meeting your own delivery promises.

Ask yourself another question. Have you ever forced a supplier to take a job he or she was not capable of just because you had no other choices available at the time? You know the approach . . . "you get all the gravy jobs, so you have to take a few of the dogs as well!" Or perhaps you had no one else who would take the job, so you gave it to a questionable supplier just so you could get it off your desk. What was the result? You've got it—disaster: poor quality, late delivery, and excess costs. Not exactly the ideal, is it? This points up why alignment is so critical.

On the other hand, if you select a supplier whose capabilities exceed your product quality requirements, a "quality gap" will develop, as illustrated in Figure 2-2. The quality gap provides you with the opportunity to enhance your product and process quality at no additional cost. Why? Because the supplier's capabilities already exist; they are part of the supplier's existing internal process capabilities and controls. Thus, the supplier is able to produce your products, to your specifications, easily and consistently with no additional processing required. And because your products fall within the supplier's normal operating methods, there are no secondary process costs. In fact, in many cases you can increase your requirements to the level of the supplier's capabilities and controls with no additional cost whatsoever. This is why quality guru Phil Crosby said, "Quality is free."

Now that you understand alignment, what suppliers will you choose?

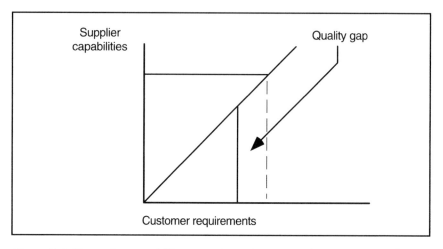

Figure 2-2. If supplier capability exceeds customer requirements, a quality gap results.

### Design and Manufacturing Process Rules

Again, let's start with a couple of questions for you to consider in setting the stage for establishing new paradigms (Table 2-3).

Obviously, the answers to both questions are no. Why, then, do we constantly have so much trouble with these two issues? The reason is often that we fail to communicate effectively. We make unfounded assumptions regarding what design and manufacturing engineers know and understand about each other's discipline and its requirements. We fail, in most cases, to develop a set of guidelines under which we mutually agree to operate. Take for example the following questions:

Question 1: Are your design engineers aware of the capabilities of your internal manufacturing processes from a statistically-based

### Table 2-3. The Design-manufacturing Dilemma

*Question 1.* Do all design engineers intentionally design products that cannot be made economically with existing tooling and processes?

*Question 2.* Are all manufacturing engineers resistant to making product and process changes that involve new tooling or modifications to existing tooling and equipment?

process capability study? If not, how do you as the manufacturing engineer expect them to correctly specify product tolerances, test requirements, reliability, and workability factors? Given no guidelines to work under, the design engineer will create "safe" tolerances and specifications that guarantee a high level of capability or robustness in the design . . . you know, those tolerances that require process capabilities that don't exist in your manufacturing processes today, or those that cost more than your normal processing techniques. And your reaction: "How could those blankety-blank design engineers possibly have come up with this? Don't they understand manufacturing?" Well, no, they don't. And you didn't educate them. End of communication; end of story.

Question 2: Are your design engineers aware of your suppliers' process capabilities from a statistically-based process capability study? Same answer? Same result. Get the buyer and the suppliers involved in the development of those studies. Then get them involved early in the new product development cycle so that specification issues can be resolved before the design is released. Remember, more than 90% of product costs are designed in. The only way to effectively reduce those costs after the design is released is to redesign the product or some of its components and subassemblies. However, this solution is neither cost-effective nor time-sensitive.

Question 3: Do you have an approved supplier listing that limits the design engineers to specifying products from only qualified or certified suppliers? If not, you may be introducing problems into the manufacturing processes in the form of nonconforming materials, late deliveries, and inadequate or excess quantities of purchased items. Stick with known capabilities and leave the gambling to the Vegas crowd.

Question 4: Has your senior management established a strategic product plan that defines both the products and the manufacturing processes and equipment required to support it? Or, put another way, do you have a 5-year business plan? The answer is typically yes. In support of that business plan, do you have a 5-year product plan that defines the products you intend to offer over the next 5 years, along with the market dynamics that define the appropriate introduction timing? Here, the affirmative responses typically fall off to 50% or less. Finally, do you have a 5-year facilities and equipment plan that defines and provides for the technology and equipment required to bring those products to market? As you might expect, fewer than 15% of American companies have planned to this extent. And yet, if we fail to plan for the future to this level of detail, the future often becomes dictated by how quickly we can respond to last-minute surprises. Too often, we address the strategic plans and forget the tactical issues until it is too late. The new manufacturing engineer must be involved in both the strategic and tactical planning of products and support processes to ensure that management's long-term business plans can be met.

Question 5: Have manufacturing and design process rules been established jointly between manufacturing and design engineers, and are they consistent with sound design for producibility, assembly, and maintainability rules? Work together to define the guidelines under which everyone must operate. It solidifies both the scope and approach of the design and development processes, thereby making everyone's job easier. Who knows, maybe together you will design the product, processes, and tooling right the first time. That makes everyone look good and delights the customer at the same time.

### Total Life-cycle Costing

Today, customers no longer buy things based solely on the initial investment or purchase price. Instead, the total cost of ownership is increasingly becoming the investment strategy in products ranging from capital equipment to cars. As manufacturing engineers, you have used the total life-cycle costing methods for purchasing capital equipment for years. Only recently, however, has that methodology been adopted by the average consumer.

To illustrate, consider recent advertisements for automobiles. In the past, the manufacturers offered standard warranty periods, originally 24 months or 24,000 miles. As competition grew, the warranty periods increased to 36 months/36,000 miles, then 60 months/60,000 miles. Today, the automotive giants are approaching the consumer from the total life-cycle perspective—no costs, only oil changes, for the first 100,000 miles. The message to the consumer? Your total life-cycle costs have been reduced, making your investment over the life of the product less. No hassles or lost time at dealer appointments for warranty work, no additional costs in transportation to and from work from the dealer, just use the product and enjoy it. Not bad logic.

How does that impact you as a manufacturing engineer? Remember this: 85-90% of the total life-cycle costs of a product are fixed at the time the design is frozen. Unless the process and quality controls are designed in at the concept stage, the cost of retrofitting existing manufacturing processes and equipment to meet new production, test, or quality requirements could make the product either unproducible or unsellable. Either way, you lose.

As a refresher, let's review the total life-cycle cost (TLC) equation.

$$TLC = AC + DC + MC + TC + TEC + IC + DC$$

where:

$AC$ = Acquisition Costs (planning costs, design costs, construction or production costs, purchase costs, spare parts costs, finance costs)

$DC$ = Distribution Costs (packaging costs, transportation costs, handling costs, warehousing costs, duties, tariffs, customs processing costs)

$MC$ = Maintenance Costs (customer service costs, planned maintenance costs, unplanned maintenance costs, unplanned downtime costs)

$TC$ = Training Costs (operator/user training, maintenance personnel training, support/sales staff training, etc.)

$TEC$ = Tool and Test Equipment Costs (diagnostics costs, special tooling costs, production tooling, handling equipment, etc.)

$IC$ = Inventory Costs (service parts, carrying costs, premium freight costs, etc.)

$DC$ = Disposal Costs (recycle costs, disassembly costs, hauling costs, landfill costs)

To be successful, every one of these cost factors must be considered in the design stage, then addressed through the product and process design to ensure that they are kept to a minimum. The manufacturing engineer must give his or her organization every competitive edge possible. Astute sales professionals have learned to "sell total cost and total quality." If they can't beat you on price alone, they will beat you with the total cost approach. Be forewarned, and be ready.

## THE CHALLENGE, THEN, IS THIS . . .

In the future, all manufacturing engineers must become more creative and risk conscious. They must be aware of, and willing to implement new design and quality confirmation techniques, manufacturing technologies, inventory and expense control methodologies, and manufacturing performance metrics in order to advance the competitiveness of their organizations. Today's manufacturing engineers are educated, well trained, experienced, intelligent, motivated, and ambitious, but too often averse to organizational thinking. What must be added to their skills are the business acumen and organizational skills required to envision the strategic changes necessary at the organizational level to make their companies, and the American manufacturing sector as a whole, more competitive in today's growing global market. That vision, or the lack of it, will set the stage for the survival or failure of the American manufacturing sector in the millenium.

## SUMMARY AND EXERCISE

A key responsibility of the manufacturing engineer has traditionally been to reduce or eliminate waste within the production operations. The real waste, however, is often a result of the inputs to the production processes: information, materials, tooling, and equipment (Figure 2-3).

When output does not meet the plan, outside factors such as labor usually get the blame—poor worker productivity, poor standards, poor work ethic, and unskilled workers and supervisors. Next are the suppliers—they never meet quality requirements, are always late, never ship the correct amount, and have excessive tooling and setup costs.

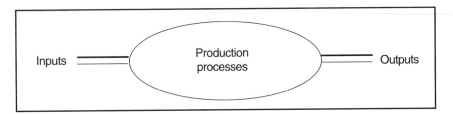

*Figure 2-3. Waste generally has its genesis upstream of production processes.*

These are all true in some cases, but to what extent? Many of the issues that introduce the greatest amount of waste into the manufacturing processes on a regular basis continue to be tolerated because the actual costs to the organization remain unknown.

In the 21st century, the manufacturing engineer must be the driver to force those hidden costs to the surface where they can be addressed. He or she must isolate and resolve all input deficiencies to the manufacturing processes to achieve the maximum productivity possible. "It's not my job" will no longer be acceptable.

Let's use a familiar component to highlight the degree to which costs are impacted in our organizations each day: the Engineering Change Notice (ECN). To drive home our point about the degree of those unrecognized costs, let's cost the ECN in your organization. Remember, every investment must have an adequate return. So it goes with every ECN. If the proposed engineering change does not provide measurable returns in excess of its cost to implement (safety issues aside) within a reasonable period, then it simply should *not* be implemented.

For this exercise, check with your cost accounting group to obtain a fully burdened hourly rate for each of the functions included in the worksheet depicted in Figure 2-4. In your analysis, be as factual as possible. Check with each function to get a good estimate of the actual time it now takes to process an ECN through each of the respective areas. Once you have a total labor cost, multiply it times the number of ECNs processed each month. That will give you an idea of the financial impact on your organization.

After you have completed the exercise, take a look at some of the typical costs calculated for several different industries by your manufacturing engineering counterparts. Compare your costs to theirs to determine how you are doing.

| Costing the Engineering Change Notice | | | |
|---|---|---|---|
| **Function or Activity** | **Hours Required** | **Cost/Hour** | **Total Cost** |
| Design engineering change | | | |
| Detailing & drafting | | | |
| Design review meeting | | | |
| Checking & approval process | | | |
| Document control | | | |
| Document distribution | | | |
| Manufacturing engineering review | | | |
| Process design change | | | |
| Tooling design change | | | |
| Quality planning & test | | | |
| Inventory planning | | | |
| Purchase of new materials | | | |
| Supplier costs exclusive of materials | | | |
| Transportation planning | | | |
| Receiving inspection | | | |
| Material handling & storage | | | |
| Material issue & movement | | | |
| Production time | | | |
| FDA, UL, ISO approvals | | | |
| Other | | | |
| TOTAL COST OF AN ECN | | | |
| NUMBER OF ECNs PER MONTH | | | |
| TOTAL MONTHLY FINANCIAL IMPACT | | | |

*Figure 2-4. Costing engineering changes reveals the full impact of design changes.*

Notice that the analysis does not include the cost of surplus or obsolete material dispositions resulting from the ECN, nor the costs associated with the conversion or scrapping of tooling and fixtures. It is only the direct administrative costs of processing each change notice—a direct drain on the profitability of the organization, never recovered, seldom budgeted, and always detrimental.

Compare your results with those reported in the industries shown in Table 2-4.

**Table 2-4. Cost of ECNs for Various Industries**

| Industry | Cost per ECN |
|---|---|
| Toys | $ 1,250.00 |
| Apparel | $ 1,750.00 |
| Pharmaceuticals | $ 9,175.00 |
| Electronics | $ 12,500.00 |
| Automotive | $ 37,250.00 |
| Aerospace | $ 65,000.00 |
| Defense | $137,400.00 |

Is there room for improvement? Does every ECN processed provide an adequate return on investment? If not, why are you doing it?

As we move into the tools developed in the following chapters that address waste reduction, constantly look for comparisons within your organization. Explore ways to begin implementing these tools and techniques immediately. Remember, time is of the essence and can be your competitive weapon.

# 3

# Concurrency

Today, a corporation can still develop a competitive market advantage through superior technology or leading-edge manufacturing capabilities. Sustaining that advantage over time, however, requires the ongoing ability to introduce new products to the market that consistently conform to dynamic customer requirements and expectations. It is those consistent new product development successes that create customer loyalty and ensure market share.

With those new products comes the requirement for new processes and equipment that address the customers' requirements for lower cost, higher quality, increased product performance and reliability, and continuous agility to meet changing markets. The effective combination of product and process development ensures that the new product development cycle is successful in meeting both its penetration and financial objectives.

Take for example how the Japanese manufacturers view new product development. The Japanese focus on gaining market share as a key element in their strategic planning. Of equal importance, however, is their belief that increasing their ratio of new to existing products not only builds new markets, but aids in further expansion of their existing customer base. Their deployment methodology is the constant enhancement of their new product and process development cycles to allow them to rapidly introduce new products, as well as enhancements to existing ones. The concept is commonly referred to as *rapid product deployment* (RPD). They are quick to point out that the RPD strategy increases their manufacturing productivity and their product quality as well, both of which contribute to reduced costs.

## REACTING TO UNPREDICTABLE
## CUSTOMER DEMANDS

Quality is secondary if your product is not introduced to the market at the proper time and at a perceived price/value point that meets customers' expectations.

Time is the competitive weapon of choice in manufacturing today and will continue to gain force well into the next century. Successful manufacturing enterprises must focus their attention on maximizing their return on corporate resources and invested time by eliminating all waste and all nonvalue-added activities within the new product development cycle. To do so, the old paradigms that focus on design parochialism must give way to a recognition of the value of a cross-functional team approach supported by a cooperative environment of concurrent functionality and worker empowerment at all levels.

Success in the manufacturing sector requires the ability to consistently produce products of superior quality, at competitive prices, (often customized individually in batches as small as one), with minimum order lead times. This necessitates agility in both the design and production phases of the new product development process, characterized by information sharing within new product development teams, within and among corporate functions, as well as with suppliers and customers. This, in turn, includes the sharing of all information relative to the product—its quality and performance criteria, target pricing, and volumes; the processes that will produce the product, including the suppliers of the raw materials used within those processes; and the applications under which the product will be used, along with the metrics that will be applied within those applications.

The concept of concurrency is fundamental to reducing new product development cycles and to the strategically planned introduction of product upgrades and enhancements. By definition, concurrency is the creation of an organizational culture in which all functional disciplines that contribute to the design, development, production, distribution, and sale of a product perform their functions simultaneously versus sequentially. The implied, underlying objective is to reduce the cycle time to its lowest possible level by designing the product along with its associated production and inspection processes at the same time.

Concurrency focuses attention in the early stages of design development where the impact of a change is less costly and has less of a detrimental impact on the total new product development cycle. Economic feasibility is confirmed early in the design stage where options are still possible, thereby eliminating the need to introduce "loss leaders" just to have a product competing on the market. Quality is designed into the product through the process and the design rules established from known production capabilities of both internal and supplier processes versus being derived from the inspection process. The result is products that can be produced, using readily available processes, at performance and cost levels that meet or exceed customer expectations.

This is in direct contrast to the traditional approach in which Design Engineering creates the product (with little, if any, thought given to its manufacturability), throws it over the wall to Manufacturing, which makes it and tosses it over another wall to Sales to peddle. With this traditional approach, the metric for design typically has been the approval of the design by fellow design engineers and managers, not necessarily that of the customer.

## THE NEW METRICS

In today's time-sensitive market, world-class organizations have re-engineered their metrics to include:
- Resource utilization,
- Design quality, and
- New product development cycle time.

Resource utilization involves the consumption or deployment of materials, capital equipment, tooling, and personnel. Under design quality are such things as the level of product producibility and performance, product reliability, total product life-cycle costs, environmental impact, and compliance with customer expectations and application requirements. Total cycle time embraces every element in the new product development cycle, including:
- Knowledge acquisition,
- Concept investigation,
- Basic design preparation,
- Prototyping and debugging/testing,
- Pilot production,

- Manufacturing ramp-up, and
- Distribution and installation.

The window of opportunity for each product, new and existing, is growing increasingly narrower. So, successes will increasingly be based on how effectively each of these metrics is achieved.

## TRADITIONAL APPROACH AND ITS COST

With the traditional approach to new product development (NPD), costs escalate as the time for each activity builds. The longer it takes to complete the cycle, obviously the more resources are consumed and, thus, more costs are incurred. That is why the traditional approach has often been so costly and risk-oriented.

Within this sequential approach, activities are performed in series, passing from one function or department to another. All departments work autonomously, causing frequent redesigns due to a lack of downstream focus and communication. In fact, the NPD focus is constantly changing as each step in the cycle is completed, because each function is concentrating on its particular piece of the pie. Information is batched, then transferred to the next functional area upon completion of the prior function's "piece." The "throw-it-over the-wall" mentality adds to the breakdown of communication throughout the NPD process, thereby ensuring that problems will surface only after it is too late to correct them without a monumental (and costly) effort. In the traditional approach, managerial effectiveness is compromised by territorial boundaries, functional priorities, and hidden agendas. Priorities are changed constantly to correspond with the pressures felt by an individual department manager on a day-to-day basis. NPD activities, because they are viewed as "long-term," are easily set aside with the lost time to be made up later. As NPD time pressures surface, the recognition that one department cannot start until it is passed the needed information from the upstream department (but will be held accountable for completing their "piece" on time) causes turf wars to break out.

## OVERLAPPING METHODOLOGIES

In an effort to compress the NPD cycle time (and cost), management gurus developed the *overlap* theory, which is based on the premise that not all activities must be 100% completed before the next activity in the sequence can start. In other words, we can safely begin the downstream activity shortly before the upstream activity is completed without undue risk of a major change causing us to start anew all of the work and time we have invested (Figure 3-1).

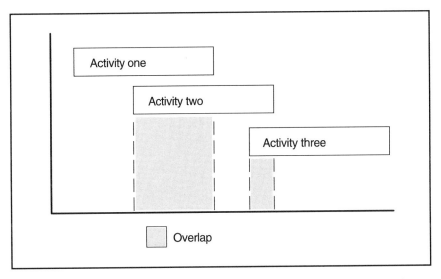

*Figure 3-1. By overlapping methodologies, manufacturing engineers can substantially compress the NPD cycle.*

As overlapping methods began to evolve, communication and interaction was improved, but only moderately. Why? Because departments would only interface with those immediately upstream and downstream of them. While information flow was better, it was by no means good. And because of the limited communication and comprehensive downstream focus, redesigns remained common. The NPD cycle remained essentially sequential, so little improvement in total NPD cycle time was realized.

Information flow continued in batch format, although small nuggets of information were passed downstream on a *need-to-know* basis. Territorial priorities continued to take precedence over those of the NPD effort, with product team members constantly changed to meet individual functional needs. There existed no "glue" to pull and keep the NPD effort together.

## WITH CONCURRENT ENGINEERING

Concurrent engineering is based on all functions, both upstream and downstream, working together throughout the entire NPD cycle—Design Engineering, Marketing, Drafting/Detailing, Manufacturing Engineering, Industrial Engineering, Tooling, Process Engineering, Materials, Purchasing, Manufacturing, Finance, Package Design, Distribution, Field Service, key suppliers, customers, and management. The NPD effort is typically led by a broad-based business executive rather than either a marketing or design engineering specialist to ensure that the total business perspective is considered. The approach is that of a true cross-functional team, responsible and accountable for results—performance, reliability, cost, configuration, delivery, and field support. Information flow is directed both upstream and downstream, processed in small batches, and transferred on line (on a real-time basis) to all team members and support personnel.

Because the team has a common purpose and focus and all members are operating from the same database, disciplines are maintained and follow-through is excellent. Problems are surfaced and resolved immediately, so significant changes are few and their impact on cost and NPD cycle time is minimal.

According to a recent Computer-Aided Logistics Society (CALS) study, examples of concurrent engineering showed:

- An 82% reduction in power-train development time,
- Number of drawings reduced from 200 to 3,
- Engineering change time reduced by 80%,
- Engineering drawing access time reduced from several days to a few minutes,
- Complete elimination of physical mockups,
- Assembly time reduced from 6 weeks to 2 hours,
- A 700% process yield improvement,
- A six-fold reduction in form and fit errors,
- A 40% reduction in inspection rejections,
- A 30% savings in inventory costs,
- An 80% reduction in rework,
- A 70% reduction in expenses,
- A 98% reduction in data transfer errors, and
- A 30% reduction in document change time.

## Why Concurrent Engineering Works

Aside from the obvious savings in NPD cycle time, concurrent engineering provides myriad other benefits to the organization:

- Concurrent engineering aids the manufacturing engineer in identifying design, process, production, and supplier problems early in the NPD cycle when they can be addressed quickly, with minimum impact on cost and cycle time.
- Promotes and enhances the development of cross-functional teams and individual skills and creativity.
- Improves company-wide communication and cooperation by breaking down functional "silos" and parochial thinking.
- Promotes "team" thinking and environment by forcing interaction at all levels and at all times throughout the NPD cycle.
- Promotes early supplier involvement for isolation of supply-base problems and capabilities prior to design lock-in.
- Allows products to be designed for existing or planned automation, thereby maximizing the utilization of available equipment and tooling by fostering group technology methodologies.
- Optimizes processes prior to design lock-in.
- Reduces the number of downstream engineering changes (and their associated costs) by forcing problem identification and resolution into the design concept and development stages. The up-front load may be heavier, but the downstream return on investment in time and total resources is significant.
- Forces the go/no-go decision point to the front of the NPD cycle where options are available, thus limiting the exposure and risk of product cost overruns and product reliability issues.
- Eliminates the "moving-target syndrome" that interferes with sound design practices, thereby giving the design engineer (and the entire concurrent engineering team) the ability to design the product correctly the first time.
- Reduces resource consumption by as much as 50-60% throughout the NPD cycle. The result is often a new product introduced to the market in half the time of the traditional approach and at less than half the cost.
- Isolates packaging, handling, and distribution problems prior to design lock-in versus when the product hits the customers' receiving docks.

The objective of concurrent engineering is simple: manage the entire NPD cycle as a process, a process that is customer-focused.

By creating an NPD process that "thinks like our customers think," we can effectively create an environment that promotes continuous improvement in both customer satisfaction and bottom-line profitability.

## ASSESSING CONCEPT AND DESIGN REQUIREMENTS

Let's again start with a few pointed questions to consider.
- Do your customers accurately and completely define all of their requirements and expectations at the time of order placement?
- Do your customers know exactly what your products will and won't do in their unique applications?
- Do your sales representatives sell within strict application guidelines or take *any* order?
- Are all affected functions involved in the product planning, development, and specifications prior to acceptance of a customer's order for a "special"?
- Are existing internal process capabilities and external supplier capabilities taken into consideration prior to acceptance of "special" orders?
- Are all support systems in place to support a new product introduction prior to its release?
  - Supply base,
  - Distribution,
  - Production,
  - Planning,
  - Costing,
  - Sales and marketing,
  - Field service,
  - Aftermarket materials and spare parts.
- Is the price point clearly identified, and is it attainable?
- Can the required production volumes per configuration be met with facilities, equipment, tooling, labor, and support staff currently available?

If the answer to any of these questions is "no," then the concept and design development stages in the NPD cycle are either incomplete or inaccurate. Remember, *before* a design is released for production, all related process and support system issues must have been identified and resolved. Manufacturing is not the place to iron out problems.

## CUSTOMER-FOCUSED QUALITY

### Total Conformance to Agreed-upon Requirements

Along with speed comes the requirement to produce a quality product, a product that conforms in every way to both the customers' requirements and expectations. The manufacturing engineer must understand that quality means:

- Defining, understanding, agreeing upon, and then meeting those agreed-upon product quality, reliability, performance, maintainability, and total cost specifications the first time, every time.
- Giving the customers the exact quantity they want—no more, no less—every time. Amortizing your long setup or changeover times with larger order quantities is not an issue for the customer to resolve, or even to consider when they place an order. As a manufacturing engineer, you must ensure that Manufacturing can produce customer-designated order quantities and be economically successful in doing so.
- Meeting customer delivery schedules, every time in the shortest possible time. Long lead times are a thing of the past. Long lead times mean a higher investment in inventory, as well as a higher exposure to the cost of changes in design or demand. Your customers are thinking JIT—so must you.
- Driving the waste out of your processes (administrative and operational) so that you can accomplish all of the foregoing objectives and still make a profit through total process control.

In short, quality in the new manufacturing paradigm means doing all of the right things right the first time!

It would seem that many former disciples of total quality managment (TQM) are now questioning its value. Why is it that *more than 70% of the TQM initiatives have ended in failure?* Wallace Company, the Houston-based winner of the Malcolm Baldrige National Quality Award, was forced into Chapter 11 bankruptcy protection.

Trade and general-circulation publications teem with articles asking if quality is dead, or if TQM is no more than a fad. Others question the motives of companies involved in quality programs— are they window dressing for customers or sincere efforts to build the best? The answer is simple. Quality is not a fad. It is not a "flavor of the month" program for management to rally the troops with or to impress the customers. It requires a total commitment, not just education, but training, along with consistent use and support. It means listening to customers to ascertain exactly what they want, then giving it to them. L.L. Bean shipped over 500,000 packages last Christmas without an error. That level of perfection would mean nothing if the service or product received was not what the customer wanted.

Quality means thoroughly understanding, then strategically focusing on the customers' expectations, and re-engineering every organizational process to ensure that those affected get it the first time, every time.

## THE BENEFITS AND CHALLENGES OF PROXIMITY

One of the key elements in enabling a concurrent product development team is collocation. While concurrent engineering practices are now recognized as vital to the development of competitive products, the adoption of those techniques is often unsuccessful. Why? Because as companies grow larger and more complex, internal hierarchical organizations (silos) evolve to manage the increased organizational size, growing technological complexity, and increased process specialization that comes with growth. In addition, company growth frequently results in the geographic dispersion of people and functions—Design Engineering in one building or location, Manufacturing Engineering in another, and so on. All of these factors inhibit many of the informal relationships that are essential in providing and fostering effective communication and cooperation.

Functional departments often focus inwardly on specific functional objectives, metrics, and priorities, creating a "silo" effect. The resulting hierarchical organizational structure, with organizational activities directed and managed by functional administrators, has therefore become ineffective in coordinating the many cross-functional activities required to support simultaneous new product and process development activities.

To offset these organizational deficiencies, many world-class enterprises are forming cross-functional NPD teams to support product conceptualization, design, development, and the transition to manufacturing. These cross-functional NPD teams provide a broader, more complete understanding of issues and alternatives because the individual team members bring with them an enterprise-wide perspective on the NPD efforts. This broad-based perspective guarantees that all critical elements of the business, from product design through distribution, are given due consideration. And, unlike the hierarchical structure of most NPD efforts, with the concurrent engineering approach the focus of the NPD team is on satisfying the external customers' product and support requirements, along with the internal customers' requirements for producibility, cost, supportability, testability, etc.

Often, concurrent NPD teams require more resources in the beginning of the NPD cycle. But because problems surface and are resolved earlier, total resource consumption is reduced over the life of the NPD cycle. A key factor in ensuring the success of the concurrent NPD team is providing the opportunity and ability for the team members to meet regularly so they may learn to interact as a real team versus just another committee. Keep in mind that the majority of communication during the NPD process is informal, and informal communication only happens when team members are in proximity to one another.

When people are placed together, with the barriers to communication eliminated, a relaxed atmosphere eventually develops, an atmosphere in which questions are freely exchanged, with differing opinions expressed without the negative influences of functional priorities or hidden agendas. Beware, however, that when you first bring the NPD team together, with all of the individual differences inherent in the members stemming from their differing experiences and perspectives, there is going to be a high degree of emotional trauma. Rest assured, it will pass quickly as group bonding begins to take place.

Collocation can take many forms. Many companies start with the concept of departmental collocation because it requires less resource commitment and is generally less of a challenge to the traditional organizational culture. Departmental collocation is based on locating complementary functions like Design and Manufacturing Engineering close together to foster interaction and to promote a better understanding of the other function's objectives, priorities, responsibilities, and activities. While still somewhat inhibiting to communication because of the physical barriers between the functional groups (walls), the improved communication and cooperation that ensues quickly begins to break down the "throw it over the wall" mentality as each function begins to understand its influence on the success of the downstream functions, and thus, the entire NPD cycle.

Team collocation, the second and typically the best approach, occurs when concurrent engineering team members are located in a single project area—no physical barriers, no outside influences, no conflicting priorities or agendas. Reporting lines are direct to the concurrent engineering team, not to functional managers. As such, the concurrent engineering team members are free to make decisions that are the best for the entire NPD effort without the fear of political reprisals.

In smaller companies where full-time resource allocation simply isn't possible, team collocation remains feasible for part-time or short-duration members, so long as they have a singular location in which they always meet, coupled with a dedicated time slot. The hours must be committed to the team by senior management, with no functional manager given the authority to override the dedicated team meeting times. For example, John is allocated to the concurrent engineering team between the hours of 8:00 a.m. and 10:30 a.m. daily. As far as John's boss is concerned, John does not exist between those hours and his work must be scheduled accordingly or redistributed to other manufacturing engineers.

Physical collocation promotes close working relationships and friendships between the team members, a factor that is essential in building a winning chemistry between the team members. A strong team chemistry enhances both the frequency and quality of communication and feedback. In addition, because of the physical proximity of the individual team members, coordination of team activities is made easier. Infrastructure requirements such as information and data networks, document distribution, and secretarial support are less demanding, thereby streamlining the NPD process by accelerating information flow.

In many large, multinational organizations, there are valid reasons why team collocation is not possible. Whenever that is the case, virtual collocation should be considered. Remember, with virtual collocation, managerial resistance and politics, as well as company size and the physical dispersion of departments and facilities, can be detrimental to success if not given due consideration and constant observance. Because of this potential, the concurrent engineering team must plan accordingly to make full use of the political advantages of a senior mentor who is in a position to break through political barriers and effectively minimize the outside influences on the team and its members.

Some of the benefits of team collocation can be achieved through such technologies as:

- Voice mail;
- Facsimile;
- Electronic mail;
- Video conferencing;
- Interlinking through groupware and shareware of personal computers and workstations via increased bandwidths and data compression techniques;
- Integrated services digital networks (ISDN);
- CAD-based solid modeling and simulation software; and
- Product data management systems that support access control, coordination, and release of data to team members at appropriate times over a network controlled by electronic signature capabilities.

## NEW PRODUCT DEVELOPMENT PLANNING

Assessment of both internal and external capabilities is critical early in the NPD cycle to allow for up-front planning of the resources required at the production and distribution end of the NPD cycle. That planning should encompass a thorough review and assessment of:

- *The master production plan.* Consideration must be given to the planned volumes per configuration; the proposed pilot and full production release dates; the scheduled material availability dates; a final confirmation of supply base readiness; practical capacity and process capability of both internal and external supply operations; confirmed tooling, fixture, and equipment availability dates; aftermarket and service parts requirements and associated availability dates; and, finally, employee readiness, along with any corresponding training schedules needed to ensure readiness.

- *Support documentation.* NPD planning should also include a thorough review and confirmation of product and process design rules; manufacturing policies, procedures, methods, routings, and costs; quality test, inspection, and final confirmation methods and procedures; audited product and assembly drawings and bills of materials; installation and application documentation; safety, warranty, and environmental impact documentation for compliance with Underwriter's Laboratories (UL), Canadian Standards Association (CSA), International Organization for Standardization (ISO), Environmental Protection Agency (EPA), or any other governing regulatory bodies; engineered cost standards; and marketing and sales literature such as price books, brochures, and advertisements.
- *Quality confirmation requirements.* Don't overlook the quality issues like critical characteristic identification and sample frequencies; Failure Mode and Effects Analysis (FMEA) and design of experiments (DOE); reliability and life-cycle testing; and any required UL, CSA, or ISO approvals.
- *Lead-time management.* Remember, *time* is the ultimate competitive weapon, so include a thorough review of both manufacturing and administrative cycle times. If the combination of the two do not conform to the required lead time expected by customers, re-engineer the appropriate direct and support processes as necessary to ensure optimum cycle times.
- *Distribution methods.* We often fail to take our final assessments far enough. Do not fail to consider distribution issues like in-house or field stocking, or floor planning methodologies; packaging designs for single unit and bulk customer orders; transportation planning that includes the preferred delivery methods and freight-on-board (FOB) point; payment of any applicable taxes, duties, tariffs, royalties, customs processing fees, etc.; internal or customer product labeling or bar coding requirements; and suggested end-user and distributor pricing.

Once the planning has been completed and the design is finalized, a limited pilot production run should be scheduled to ensure all technological and financial issues have been resolved. In addition, the pilot production run provides a final check on the production readiness, tooling and fixture confirmations, material and subassembly compatibility, and quality system effectiveness. A word of caution: pilot runs should not be used by Marketing to launch new products. Rather, the singular purpose of a pilot run is to assess the organization's readiness to efficiently produce the product. Don't be trapped into launching a product *before* all support systems have been confirmed.

In addition to in-house considerations, the pilot run provides an excellent opportunity to confirm distribution methods, test packaging and labeling systems, test market the product to confirm projected demand, and perform on-site application testing of the product in the hands of the consumer.

Only when all systems are "go," should the full production run be scheduled.

## OBSERVATIONS

In my travels around the globe, working with literally hundreds of organizations, one thing has become abundantly clear: the final product configuration (physical) rarely corresponds 100% to the engineering documentation (drawings and bills of material). As a result, the actual product costs, manufacturing cycle times, inventory management, supplier commitment, and manufacturing support and cooperation are often compromised.

It is essential that a final configuration assessment be completed prior to the scheduling of full production. The assessment should include a complete review of:

- The physical characteristics and tolerances of the product and its components,
- The product characteristics against the final engineering drawings and specifications,
- The final bills of material,
- The fabrication and assembly methods and routings,
- The operators' and owners' manuals,
- The product and packaging compatibility, and
- The product and application compatibility.

At this point, everything should be ready—the documentation, the product, producibility issues, etc. To summarize, then, prior to final release, the manufacturing engineer and his or her concurrent engineering team should ensure that all support systems are in place to guarantee a smooth transition into full production. The final support systems assessments should include the quality, manufacturing, planning, and scheduling systems, as well as procurement, personnel, and distribution systems. Aftermarket support and field service systems and facilities should also be included.

Concurrency dictates that we consider all aspects of the business that impact the total NPD cycle (Figure 3-2). In the following chapters, we look at some of the tools that allow us to effectively re-engineer the processes and methodologies we currently use to manage those business elements to ensure the greatest reduction in the total NPD cycle time.

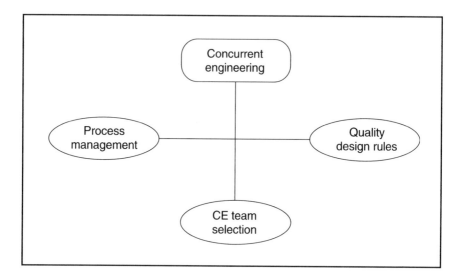

*Figure 3-2.The impact of concurrent engineering cascades to all aspects of the business enterprise.*

Before moving on, however, let's take a look at a classic example of concurrency in the government sector. That's right, the government sector. The example on the following page was extracted from an article carried in the *Dallas Morning News.*[2]

As you think through this example, consider what strategic actions the alliance deployed to reduce the total NPD cycle time. Also consider what steps you would take to introduce those same strategic actions in your organization. What would happen if your concurrent engineering team was given total freedom to do the job right the first time, without bureaucratic interference and its

## The Bunker Bomb

Thirty-seven days after U.S. Air Force officials circulated a request for help in destroying Iraqi command bunkers that had survived direct hits from 2000-pound bombs, GBU-28 eliminated Bunker No. 1 at Al-Taji Air Base. Texas Instruments (TI), Lockheed, and Eglin Air Force Base personnel created (in 37 days) a high-technology weapon of unprecedented power that would have ordinarily taken at least 18 months to develop.

On January 21, 1991, the Air Force in effect gave TI, Lockheed, and its own project staff a free hand to get the job done as quickly as possible, and worry about the paperwork and bureaucratic channels later. A Lockheed employee remembered reading about the stockpiling of old gun barrels. The barrels were located at the Letterkenny Arsenal in Pennsylvania. Within hours, Eglin personnel had a shipment of 8-inch howitzer barrels en route to the Watervliet Arsenal in New York without waiting for approval. By February 1, Army machinists at the arsenal were machining the barrels into what would become the bodies of the new bombs.

Lockheed's assignment was the warhead, TI's, the guidance systems. On February 12, before the Air Force had settled on a final bomb design, let alone contract terms, TI officials took a chance and booked the only available productive time on the Dallas-based LTV Aerospace wind tunnel, February 16 and 17. They had 4 days to build a quarter-scale model of the bomb, set it up to test body and fin configurations, and develop guidance software to achieve pinpoint accuracy with a bomb that existed only as a computer model.

The TI engineers worked in concurrent engineering teams throughout the weekend, generating what they hoped would be successful data. Monday morning they began feeding that data into computer simulations. On Thursday, at the Air Force's insistence, two completed sets of guidance systems were shipped to Eglin before TI engineers had performed their own final tests. Two more sets were requested for Nellis Air Force Base in Nevada just as soon as those tests were completed. That turned out to be Saturday, February 22.

Sunday morning, one inert bomb was test dropped on Nevada's Tonopah Test Range, where it penetrated over 100 feet into the ground. On Tuesday, the second bomb penetrated more than 22 feet of concrete at a Holloman, New Mexico test. Hours before that test, however, Eglin personnel had filled the warheads of their two bombs with explosives. They were immediately flown to Saudi Arabia where, less than 8 hours after they arrived, they were on their way to a rendezvous with Bunker No. 1. Their mission was a success.

associated approval cycles? What impact would it have on your NPD cycle time if design decisions could be made and implemented by the concurrent engineering team, without the blessing of senior management? Examples of concurrent engineering abound in almost every industry segment.

Successes like that can be yours. As manufacturing engineers, you set the stage for either successes like those cited here, or mediocrity. The choice is yours.

## CLOSING THOUGHT

"No enterprise can exist for itself alone. It ministers to some great need, it performs some great service, not for itself, but for others; or failing therein, it ceases to be profitable and ceases to exist."

*—Calvin Coolidge*

### Reference

2. *Dallas Morning News,* June 30, 1991; "The Bunker Bomb." Dallas, Texas.

# 4

# Core Process Re-engineering

**COUPLE A GOOD EMPLOYEE TO A POOR PROCESS . . .**
**AND THE PROCESS WILL PREVAIL EVERY TIME!**

Core process re-engineering is one of the hottest techniques in use today to bring everything from individual administrative or manufacturing processes to entire corporate structures up to levels that will be competitive in the 21st century. The success stories associated with the re-engineering movement are endless. Dozens of small, medium, and large companies, from start-ups to world-class organizations have re-engineered all or part of their processes to minimize cycle time and its associated costs, improve quality and customer service, and reduce unnecessary overhead costs.

Re-engineering involves a complete "clean-sheet" approach to answering the question, "If we were to start over today with nothing—no history, no procedures, no past practices, nothing—how would we structure our organization and its internal processes to ensure the best possible service to our customers?" Re-engineering requires us to rethink the way the corporation is structured and organized. For instance, many companies (most, in fact) organize work around departments or functions. Manufacturing makes the "stuff"; Finance handles the invoicing; Marketing and Sales deal with the customers; etc. That structure, as we have so often found, however, is subject to continuous errors and delays.

For example, a large nationally-known insurance company took 24 days to issue a policy, but the real work took only 10 minutes. The remainder of the time was spent waiting—either in in-baskets, or moving from desk to desk awaiting the next level of approval (sound familiar?). Re-engineering those same processes allowed the company to get their policies to the customer within 24 hours, with fewer than half the people required to shuffle the paper. Savings to the company, and its customers, were significant to say the least.

The benefits from re-engineering have been enormous to many organizations:

- Fairview-AFX, a winner of the 1995 Quality Cup Award fell into trouble in early 1992. A clerk calling customers to inquire about late invoice payments discovered a world of angry and dissatisfied customers complaining about Fairview's lack of customer service. It was the first time Fairview's senior management had a clue they were in such deep trouble.

  As complaints began to mount from even old-line customers, Fairview's management assembled a team of employees to reverse a quickly deteriorating situation. The team soon found that the cause of their customers' ire was Fairview's lack of customer service—failure to inform customers of shipment delays, excessively long equipment repair cycle times, and often just a failure to return a customer's phone call promptly. Masking the problem was tremendous business growth and excellent profits.

  Fairview's team re-engineered their customer service processes and procedures, adding "ownership" of customer service to every employee's job description. Customer service became the company's primary mission. And while there were challenges to overcome, Fairview has since seen a 76% growth in sales, with a 95% growth in earnings. In addition, customer service levels are up significantly, while accounts receivable are down. The moral? Happy customers pay their bills promptly.

- Another of the 1995 Quality Cup Award winners, K2, had converted their production facilities and manufacturing methods to produce a more popular model of snow ski, only to find that its employees were unable to produce the new ski efficiently. Scrap levels immediately skyrocketed to around 30%. And as they did, morale plummeted.

  K2, with the help of its employees, essentially re-invented its entire manufacturing and quality processes. Five quality teams met twice a week to develop standardized methods to perform critical production and quality assurance tasks. The teams documented each new method and developed new operating and quality procedures for each. Preventive measures were implemented to identify potential production and quality problems before they could occur. Raw material sourcing and supply chain management techniques were implemented to improve the quality and timeliness of materials entering the production processes.

The result—an increase in first-pass yield to 95%, a reduction in press downtime from 250 hours a week to just 2.5 hours, and a 30% increase in production output.

- A re-engineering team at Madison Paper Industries reduced process waste by 60%, saving the company over $3.25 million annually.
- IBM cut the cycle time for development of its printers from 4 years to 2. Hewlett-Packard reduced its cycle time from 54 months to less than 22 months on its printer lines.
- Northern Telecom slashed its new product development time for digital switches by 20-50%.
- Motorola introduced its pager line in just 18 months versus the usual 3-year cycle.
- Eastman Kodak reduced the time it takes to develop a 35mm camera from 70 weeks to less than 38.
- Bell Atlantic cut the connection time to long-distance carriers for its customers from 15 days to just hours, while reducing the connection costs from $88 million to just $6 million annually.
- Hallmark Cards used re-engineering to restructure the development process for new greeting cards. Using a concurrent design approach, Hallmark now empowers a team from almost every department (collocated into one room) to make the final design decisions on a new card, something formerly the realm of senior management. The result has been a 50% reduction in the total time to market from 2 years to less than 1, at a significant savings.
- Honeywell Building Products cut the NPD on its thermostats from 4 years to 1.
- Diesel truck development was slashed by Navistar from 5 years to 2 1/2 years.
- At Xerox, new copy machines are developed in just 2 years, a task that formerly took up to 4 1/2 years.
- Taco Bell approached re-engineering from an entirely different perspective—survival. In the early '80s, Taco Bell was light

years behind their competition. John E. Martin, Taco Bell's CEO put it in these words, "If something was simple, we made it complex. If it was hard, we figured out a way to make it impossible." Taco Bell's approach was to re-engineer its business from the customer's perspective, to concentrate only on those processes that added value in the eyes of the customer. At the average fast-food restaurant, there is a 4:1 markup on food and paper products, with 75 cents of every customer dollar going to marketing and overhead.

Taco Bell's re-engineering approach included:

1. Reversal of the previous floor layout from 70% kitchen and 30% customer seating to 70% customer seating and 30% kitchen.
2. Elimination of district managers, giving the restaurant managers complete profit and loss responsibility.
3. Moving cooking outside to central commissaries.

The result has allowed Taco Bell to lower its prices, while boosting serving capacity from $400 per hour to $1500 per hour.

Re-engineering results have been equally as impressive in industry segments as diverse as pharmaceuticals and aerospace. In short, core process re-engineering can be applied to any process to reduce waste and cost, while improving quality and customer service. The key point to remember is all work is a process.

## THE FIRST HURDLE

Heard any of these before?

"That works great in other industries, but it just doesn't apply to ours." "You simply don't understand my business." "Our industry is different from all the others." "Our organization is complex, our people unique, our structure proven." "That's not my job."

The simple fact is that all work, irrespective of organization, industry segment, or organizational size, is a process, a process with inputs, value-added and nonvalue-added activities, and outputs. And because all work is a process, the fundamental rules of core process re-engineering apply whether you work in a service industry, an electronics manufacturing plant, a hospital, a publishing house, or a university. By isolating and addressing those elements of the process that do not add value, or require excessive cycle time or organizational resources, the process can be improved or entirely re-engineered to provide for and facilitate the optimum use of resources in the shortest amount of time with the highest level of quality output.

The objective of the manufacturing engineer of the 21st century is simple: manage the entire business as a process, a process focused on the customer. By creating a total business process that "thinks like the customers think," the manufacturing engineer will effectively create an environment that promotes continuous improvement in both customer satisfaction and bottom-line profitability.

### Customer-focused Thinking

For an organization to be truly customer focused, every business activity and process must address the following questions:

1. Who are our customers?
2. What do we provide our customers?
3. What do our customers want?
4. How can we improve their satisfaction with our offerings and our performance?

Taking those questions internally, for each process we must ask:

1. Who is our (internal) customer?
2. What do our customers want from the outputs of our process?
3. When do they want it?
4. What do our customers do with our output?
5. Is there a difference between what we provide and what the customers expect?
6. Does our process add value in the eyes of the customer?
7. Does what we do add value to the process of satisfying the customer?
8. Does it make economical sense to continue doing what we have always done?

Henry Ford put it quite succinctly: "Our job is to satisfy our customers, and our customers are the persons our work moves to next!"

### Creating a Customer-focused Culture

Ask anyone in your organization who the most important person(s) is to the organization and they will almost universally say "the Customer." And yet, if we assess the internal processes within each organization, such as the following, there is little to support the fact that our primary objective is to satisfy the customer.

## Organizational Structure

Few organizations are structured in such a way as to:
- Promote rapid and accurate transmittal of vital customer information and requirements.
- Measure customer satisfaction, and from the results, implement corrective actions to ensure continuous improvement.
- Foster employee empowerment to enhance the process of satisfying the customer, and to promote continuous improvement of those processes.
- Emphasize organizational improvement metrics, versus rewarding on the basis of individual or functional successes.
- Meet or exceed the customers' needs, versus merely satisfying or coping with organizational politics.

## Performance Appraisal Systems

Would a review of your individual and organizational performance appraisal systems reveal systems that promote:
- Achievement of organizational versus individual objectives?
- Linkages between the mission or vision of the organization (as defined by management) to all suborganizational metrics and performance criteria?
- Rewards for team versus individual successes?
- Accountability for results and failures on a team versus an individual basis?

Have the key process issues in Table 4-1 been adequately assessed and addressed in your company to ensure you are making the right products, at the right time, and in the right quantities to meet and exceed customer requirements and expectations? If not, is your focus truly on the customer?

Remember, there are typically only three criteria for measuring customer satisfaction: (1) *Performance* (objective, quantifiable, measurable); (2) *Perception* (subjective, and only indirectly measurable); and (3) *Results* (obtained through the use of the product or service in the customers' unique application and environment).

### Customer-focused Process Management

In assessing your company's performance, remember there are two views of an organization that must be considered:
1. *The functional view*—The individual operations and performance of the separate departments or functions are viewed

Table 4-1. Assessing Customer Focus

| Key Process Requirements | Key Process Questions | Have they been Adequately Addressed (Y/N) |
|---|---|---|
| Define our outputs (knowledge, service, and product). | What do we offer? | |
| Define our customers (internal and external). | Who do we offer it to? | |
| Define our customers' expectations and prioritize them. | What do they want? When do they need it? What do they do with it? | |
| Measure the degree to which our customers' expectations are being met. | Is there a difference between what we provide and what they expect/require? | |
| Define and measure our product/service design, development, and delivery processes. | How can we improve the process performance through re-engineering? | |
| Measure and reward our results based upon customer satisfaction ratings from the customer. | How can we achieve the required culture to ensure continuous improvement? | |

as independent organizational elements, separate from the whole, and

2. *The systems view*—Identification and management of the relationships between each process step (irrespective of functional or organizational structure) to ensure seamless process performance from customer inquiry through shipment or delivery of the product or service to the customer. Purpose: to optimize management of the critical interactions between process steps which cross functional boundaries.

### The Role of Core Process Re-engineering

The role of core process re-engineering is to guide the manufacturing engineer in finding answers to the following questions for each operational and administrative step in the process. If answers are not readily available, the process must be re-engineered to ensure that those questions are addressed in the most straightforward, time-sensitive manner possible.

- Who does what?
- When is it done?
- How is it done?
- Why is it done?
- Where do the inputs come from?
- Where do the outputs go?
- What are the requirements of each?

Typical applications for core process re-engineering include cycle-time improvements (both administrative and operational); restructuring or reorganizing of processes, product lines, administrative functions (such as purchasing or order-entry), and even entire companies; enhancement of process and product quality and performance. It is also effective as an aid in employee orientation, training, and performance evaluations; for ISO 9000 quality system baselining; as an aid in problem solving and isolation of root causes; and as an adjunct to strategic planning to help isolate operational capabilities and resource requirements as they relate to the strategic issues under consideration. In addition, core process re-engineering is a valuable tool for improving interorganizational communications and enhancing interorganizational cooperation.

Core process re-engineering allows the manufacturing engineer to:

1. Link operator-level activities directly to the strategic direction of the company as established by management in the organization's mission statement;
2. Identify how work is *actually* accomplished versus how they or management think it is completed according to procedures or past experience;
3. Isolate critical functional interfaces, process waste (nonvalue-added activities), bottlenecks, time-consumers, disconnects, and information batch nodes that contribute to increased process cycle times and costs;

4. Develop tools with which to diagnose and eliminate system and process problems;
5. Isolate and significantly reduce or eliminate administrative and operational process cycle times;
6. Enhance product and service quality while lowering their costs;
7. Identify organizational and process needs in advance of requirement;
8. Guide product development efforts through time-driven process enhancements;
9. Define both process and organizational capabilities accurately;
10. Facilitate the strategic integration of organizational, functional, and individual objectives and metrics into a common set of guidelines with a single scope and purpose.

### Re-engineering Definitions

To fully understand the relationship between re-engineering and the customer, we must first agree on one or two basic issues. First, *customer expectations* are the baseline for quality in the eyes of the customer and, therefore, for the product or service provider. Determination of customer expectations during the early stages of product and service development ensures that the product or service will conform to customer requirements at the lowest total cost to both the provider/producer and the customer. Through processes like Quality Function Deployment, the "voice of the customer" is considered in the planning and development of both the processes and resources that go into the product or service provided to the end user. The customer's expectations may not always be directly measurable, but rather may require translation into precise design criteria, specifications, and performance requirements. The key is to solicit the customer's input *before* finalizing the product or service definition.

Secondly, *quality* is the degree to which all customers' expectations are met by the product or service provided. Quality is based on two factors:

1. The customers' perception of the product or service based on subjective criteria, and
2. The performance of the product or service based on measurable criteria established and agreed to by the customer and the provider.

Take, for example, the experience of a large, well-known northeastern consumer products manufacturing company. It purchases a multitude of fasteners (raw materials) from the largest, full-line hardware distributors in its area. The fasteners are used in a number of different locations in the factory and by a number of functional disciplines including Manufacturing, Service Parts, Maintenance, and Field Service. The fasteners are key to meeting the company's customer requirements for timely product deliveries and for field service support activities. As explained to the manufacturing engineering staff, senior management is disturbed because the company carries too much inventory of fasteners (2.3 annual turns), yet they frequently run short of the critically needed fasteners (7-8 times weekly), and the cycle time from the time a purchase order requisition is prepared to the time the fasteners are received is typically in excess of 30 days. The desired cycle time is 1 day. To complicate matters, the company's suppliers are upset because they don't receive payment before 90 days, even though stated payment terms are net 30 days. Moreover, the purchase orders issued by the purchasing staff contain numerous errors (actual error rate of 46%), requiring the suppliers to contact the buyers regularly to isolate the exact requirements. Also, the buyers frequently issue purchase orders that are past due when received by the supplier (24% of the time).

To address the situation, the manufacturing engineering staff developed the matrix shown in Table 4-2 to help them isolate the critical process and customer issues.

By identifying their current process requirements and performance, and comparing them against the requirements and expectations of their customers, the manufacturing engineering staff was able to re-engineer the purchasing process to reduce the total cycle time to 1 day; move to a JIT order placement policy, which increased annual inventory turns to in excess of 26; and eliminate documentation errors through electronic data capture and transfer techniques, which facilitated the payment of supplier invoices within the 30-day window.

Why was the manufacturing engineering staff assigned the re-engineering task? Because manufacturing engineers are process people, and this process was clearly out of control.

### Table 4-2. Assessing Process and Customer Issues

| Key Process Requirements | Key Issues to be Addressed | Answers |
|---|---|---|
| Define all outputs (knowledge, service, and product) | What does the Purchasing department offer? | 1. Service...providing raw materials and services to support the Manufacturing, Service Parts, Field Service, and Maintenance departments.<br><br>2. Purchase orders...to order and schedule the release of fastener deliveries. |
| Define the customers (internal and external). | Who do they offer it to? | 1. Manufacturing...to produce critical components and subassemblies for final assembly of finished goods to meet specific customer orders.<br><br>2. Maintenance...to maintain and repair critical production equipment and tooling to ensure uninterrupted production.<br><br>3. Service Parts...to support established customer order fill rate levels.<br><br>4. Field Service...to support timely installations and field repairs on customer units.<br><br>5. Suppliers...to ensure accurate and timely purchase order releases, coupled with timely payment of supplier invoices. |

**Table 4-2.** *(Continued)*

| Key Process Requirements | Key Issues to be Addressed | Answers |
|---|---|---|
| Define the customer's expectations and prioritize them. | What do they want? When do they want it? What do they do with it? | 1. Manufacturing...small lot deliveries to maximize inventory turns and reduce in-process storage and handling requirements. 100% quality to eliminate defect-caused rework. On-time delivery to support JIT manufacturing methods.<br><br>2. Service Parts...lot sizes packaged to meet customer point-of-sale quantities. High quality (99+%). On-time delivery to meet inventory turn and fill rate objectives.<br><br>3. Field Service...parts on the shelf when needed in field stocking locations or for field repair kits. No shortages and no quality-related failures.<br><br>4. Maintenance...ready availability of a variety of fasteners within 2 hours of identified need, with 95% availability off the shelf.<br><br>5. Management...no line stoppages. Inventory turns in excess of 24 annually.<br><br>6. Suppliers...accurate purchase orders. Timely payment of invoices. Continued business. |

## Table 4-2. *(Continued)*

| Key Process Requirements | Key Issues to be Addressed | Answers |
|---|---|---|
| Define the metrics you will use to measure the degree to which the customers' expectations are being met. | Is there a difference between what the purchasing department provides and what the customers expect/require? | Yes: Manufacturing, Field Service, Service Parts, Maintenance, and Management...excess inventory, wrong parts, delivery delays, and excessive lead times.<br><br>Yes: Suppliers...inaccurate POs, late POs, incomplete POs, no lead time on orders, poor payment record. |
| Define the current and targeted product/service creation and delivery process cycle time. | Is there a gap? If so, how will it be addressed? | 1. Assess and re-engineer the material requisition, planning, and execution process based upon internal and external VOC requirements and priorities.<br><br>2. Design into the re-engineered purchasing process the internal controls to ensure PO data is complete and accurate prior to release, utilizing EDI to reduce order cycle time.<br><br>3. Install performance metrics to ensure improvement continues.<br><br>4. Develop and implement a supplier certification process to improve external quality and performance, while reducing external lead times and costs. |

Table 4-2. *(Continued)*

| Key Process Requirements | Key Issues to be Addressed | Answers |
|---|---|---|
| | | 5. Redefine packaging requirements in line with JIT and *kanban* levels for Manufacturing and customer order quantity levels for Service Parts. |
| | | 6. Pursue supplier stocking programs, placing the responsibility for delivery and line stocking in the hands of the suppliers. |
| | | 7. Implement bar coding to facilitate receiving activities and in-process cycle counting transactions. |
| Define how you will achieve desired results as perceived and measured by the customers and how you will reward success. | How can the Purchasing department (and any other process contributor) achieve the desired cultural change and maintain it? | 1. Re-engineer the purchasing evaluation and incentive systems to comply with the organizational level performance requirements. |
| | | 2. Provide ongoing training in required purchasing techniques based upon a skills assessment of department personnel. |

### Where to Begin

Planning is the first leg of the critical path; it is essential to success. *Those who fail to plan, plan to fail.*

The following step-by-step procedure details the re-engineering methodologies used today by manufacturing engineers in many successful world-class companies to address quality, performance, and cycle-time deficiencies within both operational and administrative areas.

*Step One: Identify the Critical Business Issues*
The first step in improving your organization's process performance is to understand clearly and accurately which issues are critical to the success, and sometimes survival, of the organization (as defined by the customer). *Critical business issues are measurable goals or initiatives identified by senior management at the organizational or business level, based on a current or potential market (customer) problem or opportunity, that have a significant impact on the strength and sustainability of the organization.* These issues might include product life-cycle costs, product performance, reliability, timeliness of service or product delivery, service support, user friendliness, safety, environmental impact, responsiveness to market changes, product quality, etc.

Remember, the priorities of a business culture are reflected in both what is measured and how it is measured, *and* how the results of those measurements reflect on the risks and rewards realized by everyone at every level in the organization.

*Step Two: Understanding the Processes Involved*
*That which is not measured will not improve.* Once the critical business issues have been isolated, the next step is to develop a precise understanding of the processes which create, contribute to, impact or influence (directly and indirectly) those issues. Those processes will seldom be unifunctional in either design or operation. Rather, they will be broadly cross-functional and multi-influenced, requiring the assessment and critique of those process owners both upstream and downstream of the targeted process. It is only through a comprehensive assessment of all of the "input processes" along with both the "direct and user processes" that the total business enterprise can be optimized and the critical business issue satisfied. As such, it is essential that those process owners that have the greatest knowledge of and control over the targeted business process(es) be involved in the improvement/correction activities. The manufacturing engineer and his or her re-engineering team must be given clear objectives and accurate metrics with which to operate, metrics that are determined at the organizational level as opposed to those established at the individual or functional level.

Most manufacturing engineers pride themselves on their understanding of how things operate. The truth is, however, that most processes are not completely understood by anyone at any level. They have grown too complex, been unofficially modified or redesigned with little or no supporting documentation, or simply have been dropped by operators who found them to be cumbersome and ineffective. The bottom line is, the processes are out of control—out of the control of the process owner, out of the control of management. The peculiar thing is, the process customer knows the process is out of control. Why? Because the output of that process fails to meet his or her stated expectations and requirements.

It is, thus, critical that the "as-is" baseline be determined as accurately as possible to ensure the manufacturing engineer and the process re-engineering team begin with the correct information and assumptions with which to re-engineer the process. Neither the manufacturing engineer nor the team can assume they know how the process works, nor assume it is operating under existing policies and procedures. They must determine *how it actually is operating and what metrics are in place to monitor the performance of the process and the compliance of its outputs.* The resolution of the baseline determination will often take several iterations and require many detailed discussions before it is completed. But it is vital that it be right before the re-engineering activities begin.

*Step Three: Identifying the Three Levels of Performance in Your Organization*
Within every organization, there are three distinct levels of performance: the organizational level, the process level, and the individual level. To fully understand the differences, we'll take a look at each level of performance.

**Performance at the organizational level.** At this level, performance is judged as an aggregate of all internal processes and systems within the organization. The metric is the product or service that the organization provides to its customers. Variables that impact performance at the organizational level include: corporate vision, management's strategic direction and mission, management-set organizational initiatives and metrics, the organization's structure and political environment, allocation of available resources and their deployment, and the organization's capabilities both technologically and operationally to compete in the targeted markets.

Critical questions that must be answered at the organizational level typically include:
- Has the organization created a vision for its future?
- What is the organization's strategy and direction?
- Does the strategy and direction make sense in terms of the vision?
- What are the organization's expected outputs?
- Are all required functions in place to support those outputs?
- Are there any redundancies?
- Are the current processes adequate? Appropriate?
- Is the current organizational structure appropriate?
- Are organizational objectives and metrics correctly focused? Accurate? Communicated? Understood?
- Are the required resources allocated and deployed?
- Are the critical interfaces between departments being effectively managed to ensure nothing "drops through the cracks"?

**Performance at the process level.** At this level, performance is judged by the results achieved against the *process metrics* by all functions that contribute to a given process in the aggregate. For example, all functions or departments that play a part in the order entry process would be judged against the metrics established for the order entry process versus those established for or by the individual departments or functions (if they were different) for themselves as a separate entity. Variables that impact performance at the process level include linkage of process goals to customer and organizational objectives, expectations, and requirements; effectiveness of the process in obtaining the established metrics at the lowest total cost to the organization; effective and efficient management of the process to ensure process control and capability are maintained; and allocation and deployment of the required resources to support and maintain the process on an ongoing basis.

Critical issues at the process level include:
- The fact that cross-functional processes are seldom managed effectively because of conflicting functional priorities and objectives.
- Identification of critical business issues and associated critical processes that impact performance criteria established by and for the customer.

- Accurate analysis of the process, coupled with development of effective corrective actions that lead to sound re-engineering methodologies and desired results.
- Timely and comprehensive implementation of the re-engineering plans, including a critical evaluation of the results obtained.
- Continual monitoring and maintenance of the process to ensure control and continuous improvement.
- Establishment of measurable continuous improvement goals for the process (i.e., 10 times improvement in cycle time every 5 years, 10 times improvement in quality every 2 years, and so forth).

**Performance at the individual level.** At the individual level, performance is judged on the employee's contribution to the process objectives and metrics, as defined in his or her individual job description and responsibilities. In the world-class environment, the individual metrics are derived from the process metrics, which are in turn derived from the organizational level metrics (which come from the customer).

Variables that impact performance at this level include individual skill sets and training, communications and feedback systems, politics and functional priorities, task interference, conflicting direction and objectives, and work process definition. The key to success at this level is to link the individuals' responsibilities and performance metrics to the process level requirements, effectively closing the loop.

Critical issues at the individual level include:

- The availability of a clear definition of performance requirements and expectations, a set of consistent priorities and direction, and the allocation and deployment of the required resources to perform the assigned task.
- The need for trained individuals with the necessary skill sets, capabilities, capacities, drive, and dedication to accomplish process and performance objectives that have been established.
- The requirement for results that are measurable and that conform to the established customer metrics. Process standards must be attainable and consistent with levels that are supportable by existing time and resource constraints.
- The support of a reward system that promotes the positives while minimizing the negatives. Incentives must be both tan-

gible and of recognizable value to the individual. There must be, however, recognized accountability for meeting performance standards and customer expectations at this level.
- The existence of feedback systems that provide immediate, accurate, continuous, and relevant information to the individual.

*Step Four: Determining What is and What is not Value-added*
The next step in the re-engineering process is for the manufacturing engineer to define for his or her organization—again as specified by the customers—those activities and functions that add value to the service or products offered to his or her customers (internal and external). To do so, the manufacturing engineer must have a clear understanding of what value is.

*Value-added activities* are those which contribute to the strategic direction and the strategic initiatives of the organization in the most cost- and time-effective manner possible. Value-added activities do not, in any way, contribute to the suboptimization of a process or organization for the benefit of individuals or individual functions within the organization.

*Nonvalue-added activities*, conversely, may appear necessary when viewed functionally or individually, but do not contribute to the process of satisfying the customer when viewed from the process or organizational levels. Nonvalue-added activities frequently optimize an individual or functional goal at the expense of a process or organizational goal, are not cost- or time-effective, and are typified by rework, rejection, excessive costs, and cycle-time extensions.

*Step Five: Determining the Cycle-time Impact*
The overall objective of the manufacturing engineer is to determine the "as-is" cycle-time baseline of the current process, then to re-engineer the process to eliminate, or significantly reduce, those elements in the process that generate nonvalue-added time. The key is product and process simplification to reduce complexity through the elimination of time consumers, bottlenecks, process disconnects, unnecessary operations, distance, materials, and labor which do not add value to the customer or the process of satisfying the customer.

The methodology commonly used for this step is process mapping—identifying each element of the existing process; assigning a process owner, supplier, and customer to that element; and determining the time each element consumes within the total process cycle. The result is known as the "as-is map" and becomes the baseline for measurement of future improvement.

**Measuring cycle time.** Cycle time is simply the total consumed time from the beginning of a process step to the end, or from the beginning of a process to the end. It involves all segments of time, including breaks, wait time, etc. Typically calculated in cycle time are:

- Equipment downtime (planned and unplanned);
- Overtime;
- Waiting time (for information, tools, data, equipment, process);
- Queue time;
- Setup time;
- Rework, reprocessing, or reorganizing time;
- Information delays;
- Material delays;
- Administrative delays;
- Approval delays;
- Paperwork, information, or product backlogs;
- Copy making;
- File maintenance;
- Coffee breaks;
- Chats over a cigarette;
- Distances traversed;
- Inspections;
- Equipment warmups and cool-downs;
- Cleanup time;
- Time spent looking for tools, files, figures, data—anything; and
- Meetings.

In short, every minute of consumed (and wasted) time from the beginning of the process until it is concluded.

*Step Six: Identifying the Process Owners*

There is no question that change is difficult. It is, however, even more difficult when you are assigned the task of re-engineering a process under the control of another function or individual. A natural resistance to "protect one's turf" will invariably arise. There-fore, the best approach is to involve those *process owners* in the re-engineering process as part of your process re-engineering team. Process owners:

- Are typically senior managers;
- Typically possess a general understanding of the entire pro-cess, with specific knowledge regarding a number of the indi-vidual operations within the process;
- Have the ability to influence the decisions that impact the pro-cess, its inputs, and its outputs;
- Usually manage a large number of people working within the process;
- Have much to gain or lose from the success or failure of the process;
- Are held responsible and accountable for the process and its outputs;
- Generally understand the internal and external market and business environments, as well as their positive and negative impact on the process, and vice versa.

*Assessing Risk Factors*

The active involvement of the process owners, versus their ac-tive resistance, will add greatly to the success of the process re-engineering effort. But how do you know if the process owners and the individuals operating within the processes are open to change? Let's take a quick assessment of the risk factors associ-ated with the re-engineering process in your organization by com-pleting a simple exercise.

Answer the questions in Table 4-3 relative to your company, its culture, its people, and the processes you believe to be most in need of re-engineering. Assign each question a score of 1-5 using the following criteria:

1 = I completely agree with this statement.

2 = I generally agree with this statement.

3 = I neither agree nor disagree with this statement.

4 = I disagree with this statement.

5 = I adamantly disagree with this statement.

Once you have assigned a score to the individual questions, sum the scores for an overall total.

## Table 4-3. Validating the Re-engineering Process

| Question on Process Re-engineering | Scoring for this Question |
|---|---|
| 1. The purpose of and the reasons for implementing process re-engineering will be easily understood by the process owners and their people. | |
| 2. In the eyes of the internal and external customers, changes to our processes are necessary. | |
| 3. The appropriate people will be involved in or kept apprised of the re-engineering planning by our management. | |
| 4. Communication regarding core process re-engineering and its objectives will be adequate throughout the organization. | |
| 5. I believe that the changes introduced through core process re-engineering will have a minimal political impact on the people involved. | |
| 6. I believe there will be tangible rewards associated with the re-engineering process outcomes. | |
| 7. I believe the changes resulting from the re-engineering process will support the values and visions of our organization. | |
| 8. I believe there is strong management commitment to the re-engineering process. | |
| 9. I believe the relationships between individuals, functions, and departments will be enhanced as a result of the re-engineering process. | |
| 10. I believe the required resources necessary to support and implement re-engineering will be made available. | |

## Table 4-3. *(Continued)*

| Question on Process Re-engineering | Scoring for this Question |
|---|---|
| 11. I anticipate the company will realize a positive financial return from this re-engineering process. | |
| 12. I believe that an appropriate amount of time will be allowed for the implementation of the re-engineered process changes. | |
| 13. I believe that the workload of the individuals affected by the re-engineering process will be given due consideration in the planning, execution, and implementation stages. | |
| 14. I believe my position and function will be impacted positively by the re-engineering process. | |
| 15. I support the re-engineering process fully. | |
| 16. I believe employees will not be harshly treated for making an error during the implementation of the re-engineering process and the subsequent changes mandated by it. | |
| 17. I feel secure about the manner in which I will conduct my responsibilities as a result of the re-engineering process changes. | |
| 18. I believe the people responsible for the re-engineering initiative have the requisite skills to implement the changes in my area, or will be adequately trained to do so prior to initiation of the re-engineering process. | |
| 19. I trust the individuals responsible for the implementation of the re-engineering process. | |
| 20. I trust the sponsors of the re-engineering process. | |
| 21. The re-engineering process will neither overly stress nor overly burden me. | |

## Table 4-3. (Continued)

| Question on Process Re-engineering | Scoring for this Question |
|---|---|
| 22. I am not threatened in any way by the core process re-engineering initiative. | |
| 23. My personal goals are compatible with this re-engineering process and the expected outcomes. | |
| 24. It will be easy to reverse the re-engineering process if it does not satisfy all of our needs. | |
| 25. I do not believe this re-engineering change process will impact me negatively based upon my past performance. | |
| Total Score: _____ points | |

How did you do?

A score of 61 or higher indicates a significant level of resistance to the core process re-engineering implementation. As such, the risk of failure is high both during and after the implementation process. To improve your chances of success:

1. Decrease the scope of the project. You may be looking at a system versus a process, or you may not have allocated sufficient resources to get the job completed as scheduled.
2. Increase both the resources allotted and the time allocated to the implementation process.
3. Expand the training in pockets of significant resistance.
4. Change some of the players as a last resort.

A score of 35 to 60 indicates the potential of moderate resistance both during and after the implementation process. In this case, especially when the score is closer to the lower level, training concentrated in the pockets of maximum resistance will usually handle the problem and bring the level of risk into acceptable limits.

A score of 34 or less is optimum, as it indicates a low level of risk resulting from a low level of resistance throughout the organization. The organizations that score in this range are ready, willing, and able to undertake the change processes associated with re-engineering.

*Step Seven: Developing the Customer-supplier Metrics*
As we discussed earlier, an organization that establishes customer-focused metrics as a foundation for its re-engineering process changes is one that really cares about satisfying its customers. The problem, however, is that many companies establish their metrics based upon what they "think or interpret" the customer's priorities and requirements are, rather than actually getting the *voice of the customer* (VOC) to guide them. Take, for example, the following comparison of the priorities of a typical manufacturer versus those of a typical customer.

| **Manufacturer's Priorities** | | **Customer's Priorities** |
|---|---|---|
| Cost to provide | vs. | Cost to own or use |
| Work to a fixed schedule | vs. | Minimum lead times or notice |
| Provide in volume | vs. | Buy in minimum lot sizes |
| Standardization | vs. | Variety/customization |

The differences, if misunderstood, can be disastrous. So how do we accumulate VOC information? First, don't listen to your sales representatives unless they have data to support their position. (Remember: In God we trust, everyone else brings data!) This is not to say that the sales organization has been shooting in the dark. It means, rather, that markets are dynamic. What was acceptable last year is likely to be totally unacceptable this year.

Some of the more commonly used data collection techniques include:
- Surveys,
- Focus groups,
- Direct observations,
- Interviews, and
- Questionnaires.

The key is to remember that all data must be validated, then converted into usable information through statistical techniques or commonly-used methodologies like Pareto analyses, histograms, run charts, etc. Never assume the data you have collected in the raw state provides a comprehensive insight into the customers' requirements or expectations of your products or services.

**Developing metrics for customer-focused re-engineering.** Identifying the needs, expectations, and priorities of the customer is accomplished through the use of an affinity table (Figure 4-1) which structures the data collected using the techniques mentioned earlier into usable information. The steps used to develop the affinity table are enumerated in the following paragraph.

1. Define the specific process, product, or service designated for the re-engineering focus.
2. Identify the customers, internal and external.
3. List the VOC statements exactly as they were stated by the customer from your data collection activities (focus groups, interviews, observations, etc.) on a separate sheet of paper.
4. Organize the VOC statements into logical categories for grouping. For example: order-entry cycle time, purchase order accuracy, product development cycle time, etc.
5. Identify possible quality and performance metrics that support the VOC statements. Use questions like the following to guide your brainstorming efforts:

    How much _____ ?
    How many _____ ?
    How often _____ ?
    How long _____ ?

    For example, if the focus is on product development cycle time, "how much" could refer to the time spent currently and the ideal or targeted time. "How many" could refer to the number of design changes that occur throughout the entire product development cycle that delay completion. "How often" could refer to the frequency of customer-requested changes and/or the frequency of sales-requested changes. "How long" could refer to the length of time it takes on average to complete each change.

6. Determine the relationship between each metric and each VOC and record it on the matrix with an $X$ or an $O$. An $X$ indicates either a direct or inverse relationship exists between the VOC and the metric. An $O$ indicates that there is no relationship between the VOC and the metric.

**Process Metrics for Customer-focused Re-engineering (EXAMPLE)**

Product, Process, or Service: 1/3 HP 115V Motor
Customers (External): Service Parts Department
Customer (Internal): Final Assembly Department

| VOC Categories | Voice-of-the-customer Statements | Quality and Performance Metrics | | | | |
| --- | --- | --- | --- | --- | --- | --- |
| | | Final Assembly Cycle Time | Total Life-cycle Cost | Product Performance in Field | Zero Defect Quality | On-time Delivery |
| Quality | Must work the first time and every time | X | X | X | X | X |
| | Coatings must be consistent | X | X | X | X | X |
| | Must operate quietly—under 87dB | X | O | X | X | X |
| | Must conform to UL, CSA, and ISO restrictions | O | O | X | X | X |
| | Must be free of vibration | X | X | X | X | X |
| Reliability | Bearing life must exceed 25,000 hours | O | X | X | X | O |
| | Must operate in humid or wet conditions (90%+ humidity) | O | X | X | X | O |
| | Must operate in dusty conditions without premature failure | O | X | X | X | O |
| Lead Time | Want availability within 2 days for normal service orders | X | O | X | O | X |
| | Must conform to kanban pulls of 2 hours for production | X | O | O | O | X |
| | Special orders available within 5 days of order | X | O | X | O | X |
| | **Totals** | 7 | 6 | 10 | 8 | 8 |
| | **Priority** | C | D | A | B | B |
| | **Target Values** | 2 hr | $575.00 | 99% Up | 99.90% | 99.50% |
| | **Data Collection Methodology** | Time Studies | Cust Survey | Life-cycle Test | SPC | Receipt Date |

*The Consulting Alliance Group*

*Figure 4-1. Structuring process metrics helps to translate data into usable information.*

7. Total and record the total number of direct and inverse responses for each metric. The higher the number of those responses, the more relevant the metric is to the customer.
8. Prioritize the metrics utilizing a standard Pareto analysis to determine which to address first.
9. Determine a benchmark value for each selected metric that is consistent with best-in-class performance and customer expectations, as well as your corporate mission. Always consider your objectives from the *customer's* perspective.
10. Define the data collection and reporting methodologies you will use for each metric and validate the collection systems before initiating their use.

*Step Eight: Re-engineering the Process for Improvement*
With (1) the critical business issues identified, (2) the associated processes assessed for cycle time and value-added/nonvalue-added activities and functions, (3) the process owners identified and their support and active involvement obtained, and (4) the customer-driven metrics defined, it is time for the manufacturing engineer and his or her re-engineering team to begin the re-engineering processes. Remember, time can be a tremendous ally or a serious enemy. As the process times are compressed, quality goes up, costs go down, waste is eliminated, and risks are reduced. Reaction to dynamic customer requirements and expectations is enhanced because the time-focused process is both flexible and agile, a winning combination.

**The Process Management Worksheet.** Figure 4-2 shows an example of the Process Management Worksheet, which provides the guideline for your re-engineering work efforts. As a tool, its value is implicit in the structuring of your "as-is" process map to ensure that each of the critical elements in the existing process is considered. Shortcuts can be disastrous, so use the worksheet as a guide to take you efficiently from your "as-is" conditions to your "should-be" targets.

### Mapping the Process: The "As-is" Map

*If we continue doing things the way we have always done them, we will continue to get diminishing results until competition drives us out of business!*

As discussed earlier in this section, those issues most critical to the success and sustainability of the organization must receive the

## Process Management Worksheet

*The Consulting Alliance Group*

Re-engineering Project Description:

Critical Processes Involved:

**Process Step**

| As-is Cycle Time | Target Cycle Time | Gap (if any) | Root Cause of Gap | Inputs to Process Step | Process Step Outputs | Process Step Metrics | Process Step Owners | Corrective Action Plans Inititated | Re-engineer Results (Time) |
|---|---|---|---|---|---|---|---|---|---|
| | | | | | | | | | |
| | | | | | | | | | |
| | | | | | | | | | |
| | | | | | | | | | |

**Process Step**

| As-is Cycle Time | Target Cycle Time | Gap (if any) | Root Cause of Gap | Inputs to Process Step | Process Step Outputs | Process Step Metrics | Process Step Owners | Corrective Action Plans Inititated | Re-engineer Results (Time) |
|---|---|---|---|---|---|---|---|---|---|
| | | | | | | | | | |
| | | | | | | | | | |
| | | | | | | | | | |
| | | | | | | | | | |

**Process Step**

| As-is Cycle Time | Target Cycle Time | Gap (if any) | Root Cause of Gap | Inputs to Process Step | Process Step Outputs | Process Step Metrics | Process Step Owners | Corrective Action Plans Inititated | Re-engineer Results (Time) |
|---|---|---|---|---|---|---|---|---|---|
| | | | | | | | | | |
| | | | | | | | | | |
| | | | | | | | | | |
| | | | | | | | | | |

*Figure 4-2. A matrix worksheet simplifies the task of ensuring that the re-engineering process encompasses all critical elements.*

initial focus. These issues are identified through a VOC analysis at the organizational level. Senior management then establishes the overall thrust of the re-engineering activities at the process level based on the customers' priorities identified in the VOC analysis coupled with the strategic direction established by senior management for the organization.

To develop our skills, let's use a sample case study taken from a real-world situation. The principles you apply to this exercise can be applied equally as well to your own organization, be it a product or service-based industry. As you progress through this case study, constantly ask yourself, "how can I apply what I am learning to my own situation?"

---

### *Alliance Consumer Electronics Group, Inc.*

Alliance Consumer Electronics Group, Inc. is a division of AGI headquartered in Kansas City, Missouri, with service and distribution centers in Seattle, Wash.; San Diego, Calif.; Raleigh, N.C.; New York, N.Y.; and Detroit, Mich. ACEG provides repair and aftermarket parts services for each of the 14 lines of home electronics products produced by the parent company, AGI (stereo equipment, VCRs, home and automobile security systems, small appliances, and computer hardware, including laser and ink-jet printers).

Management of ACEG has conducted a VOC analysis of its customers and isolated a significant problem that has caused a loss in market share for the parent company of 12.7% in the last 2 years: the return-repair process for customer products averages between 45 and 50 days. Customer requirements dictate a total cycle time of less than 5 days, with expectations of less than 3 days.

The following is an explanation of how customer returns are processed, as identified by the employees of ACEG.

When a product breaks down in the field, the customer is required to call ACEG's 800 hotline for a Returned-Goods Authorization (RGA). A customer service representative handles the call and determines if the return is authorized and if it should be a warranty or nonwarranty item. A number is given to the customer from the RGA log maintained by the service representative, along with the address of the service center he or she is to return the product to for repair. This normally takes 10-15 minutes. After the customer has hung up, an RGA form is prepared (manually) with the appropriate customer, product, and requested service information, a 5-minute process. The service representative then sends the RGA form to his or her supervisor for approval, with a typical turnaround of 4-8 hours.

Once the signed RGA form is returned to the service representative, the data is entered into the computer system and six copies are generated for distribution, a 10-minute process. Copies are distributed to Customer Service, the Repair department, Accounting, Receiving (where they are filed for later referral, a 2-3-minute process in each department), and to the customer. When parts arrive from the customer, they are delivered directly to the Repair Department after the RGA number is confirmed and matched against the RGA in the receiving file, a 15-minute process. If an RGA has not been issued, the receiving clerk places the customer's products in a waiting area and prepares an Unauthorized Customer Return Form in duplicate that identifies the customer, customer location, product model and serial number, and requested repairs. One copy is mailed to the service representative for the customer's region and one copy is maintained in the receiving files, 25-30 minutes for the receiving process. Owing to workload, it is a full 36-48 hours before the service representative handles the paper and takes the next action. Upon review of the Unauthorized Customer Return Form, the service representative contacts the customer to isolate the source of the problem and determine what repairs are being requested. An RGA form is then prepared and processed as before: total process time approximately 45 minutes if the customer can be reached on the first try. Average time, however, is 8 hours. In addition, the service representative prints and mails a written warning to the customer regarding the RGA policy in effect, another 15 minutes. Upon receipt of the RGA in the mail from the customer service representative (8-12 hours on average), the receiving clerk relocates the products in the waiting area, matches the RGA copy against the customer's packing slip, updates his or her RGA log, prepares a receiving report, and forwards the customer's products to the Repair department. Copies of the receiving report are distributed to the Customer Service department, Accounting, the Repair department (with the customer's products), with a copy retained in Receiving. Because of the cramped space in the Receiving department (and the heavy workload), parts often are misplaced, resulting in an average time to complete this step of the process of 7-10 days.

Upon receipt of the receiving report, the Customer Service department clerk prepares a Sales Order for internal processing of chargeable and nonchargeable costs, a 15-20-minute process. Eight copies of the Sales Order are prepared, then distributed to AGI'S Manufacturing, Quality, and Engineering departments, and the Shipping, Accounting, and Repair departments at ACEG, with two copies retained in the Customer Service department for filing in the customer file and the product file, another 30-minute process.

When the customer's product reaches the Repair department, inspection and tests are completed to determine the source of the problem and if warranty or nonwarranty repairs are required. Any damage that will void the warranty is photographed for retention in the job folder, which takes 8-16 hours depending upon workload. A job folder is then prepared with a Returned Goods Report (RGR) detailing the required repairs, the RGA form, the Sales Order form, and any photographs taken. If nonwarranty repairs are required, but not approved on the RGA, the job folder is returned to the customer service representative for resolution with the customer. When the customer service representative receives approval from the customer for the repairs and an account to charge the repairs against, the job folder, Sales Order, and RGA form are updated and returned to the Repair department, a 4-6-day process.

The Repair department then repairs the customer's products and notes the associated costs on the Sales Order form. When the product repairs are completed and tested, the job folder is updated, the Sales Order form is updated with the material and labor costs, and the RGR is completed with the test results and actions taken to repair the customer's products—15-20 days for the repairs, 2-3 days to complete the documentation. The customer's product is then sent to Shipping with a copy of the RGR and the RGA form. A copy of the Sales Order form, with the repair costs, is sent to Accounting for invoicing, and a second copy is sent to the Customer Service department for filing. The job folder is then filed in the Repair department for future reference, another 30 minutes.

Upon receipt of the Sales Order form, Accounting prepares a customer invoice for nonwarranty repairs. The charge is discussed with the customer service representative for final determination of whether or not to charge the customer for the repairs: 30 minutes. If it is determined to charge the customer, an invoice is prepared and sent to Shipping for inclusion with the customer's products, another 15 minutes. If it is determined not to charge the customer, a "no charge" invoice is prepared and sent to Shipping for inclusion with the customer's products. In either case, two copies of the invoice are sent to the Customer Service department for filing in the customer and product files: another 15 minutes.

Upon receipt of the customer's products, RGR, RGA form, and the invoice from Accounting, the shipping clerk prepares the packing slip, bill-of-lading, and freight invoice; packs the customer's products in a suitable container, and notifies the carrier for pickup, all of which takes 30-45 minutes. The RGA, RGR, and shipping documentation are filed in the Shipping department, with copies sent to Customer Service for their files: 10 minutes or more.

Customer Service, upon receipt of the documentation from Shipping, updates the customer and product files, and closes the Sales Order, another 30-minute task.

*Exercise*

Based upon this information, develop an "as-is" map of the returned goods process of ACEG. Upon completion of the map, re-engineer ACEG's process to achieve the maximum reduction in cycle time without negatively impacting customer service. Be radical in your thinking. Don't let past or existing paradigms hinder your creativity. For your reference, the solution to this exercise is included in the back of this chapter. The following hints may assist you in your re-engineering efforts.

### Developing the Cycle-time Model

The objective of the development of an "as-is" map is to identify exactly how a given process works. In the example, the process structure was well defined. In your process re-engineering initiatives, however, the process structure may or may not be known accurately. As such, each step of the process must be validated, each input confirmed, and each action and output reviewed to ensure the entire process is accurately defined. It is only when we know the baseline of our processes that we can make decisions relative to the improvement of same. This is where the Process Management Worksheet comes into use.

Once the process map is completed and verified for accuracy, the manufacturing engineer and his or her re-engineering team should:

- Determine the time required to complete each step of the process with a goal of reducing time wherever possible;
- Consider the distance between each process step and the time element associated with each;
- Evaluate the flow of the process, i.e., does each step flow from the previous step to the subsequent step in a logical manner? Does the flow make good sense from a time perspective?
- Determine if there are places in the process where information, materials, or activities become bottlenecked or queued for batching;
- Determine if there are places in the process flow that appear to be disconnected from either the preceding or subsequent steps in the process;
- Determine if each step in the process adds value
  - to the process,
  - to the customer,
  - to the organization;

- Determine if any of the steps can be eliminated altogether or incorporated into another step; and
- Determine if the setup times for each step of the process have been optimized.

In many cases, the re-engineering effort will immediately illuminate *islands of discontinuity:* process steps, or entire processes, that just do not make good sense. In most cases, these are processes or procedures that have been carried down from year to year for no known reason other than "we've always done it this way." When a fresh perspective (the customer's) is brought in to review the processes, these old methodologies are found to be no longer necessary, or even logical. It is not uncommon for the re-engineering effort to yield a 50-60% reduction in process cycle time by just eliminating these redundant or unnecessary steps.

*Metrics*
For each step of the process, the manufacturing engineer should determine:

1. If a performance metric (or metrics) exists.
2. If it does, does it comply with the metrics at the organizational level as established by senior management in conjunction with the VOC?
3. Are the process metrics effectively managed and controlled?
4. Are the metrics understood by those most directly impacted by them and who have the greatest influence on the process' compliance with those metrics?
5. Are the metrics "time-focused" and results-oriented?
6. Are the processes capable of attaining the metrics established for them?

If any of these are found in the negative, it must be re-engineered along with process steps it is associated with. Remember, work toward simplicity. Unnecessary complexity creates problems, wastes time, reduces process output and quality, and frustrates the process owners.

*Which Mapping Technique Should You Use?*
In developing the "as-is" map, two techniques are frequently used. Each is effective in pictorially defining the "as-is" process, and each is compatible with the re-engineering process. The fundamental

difference is the user's preference and comfort in using one technique versus the other.

**Organizational mapping.** This is the process of pictorially displaying an organization, system, subsystem, or process from the horizontal or functional view as opposed to the classical silo or vertical view as typically denoted in an organizational chart. Organizational mapping provides an excellent macro view of the process and, as such, is a good starting point from which to gain a general understanding of how the process works.

**Process mapping.** This is a visual "activity flow" methodology that illustrates in a more detailed manner how a process flows from element to element versus from one organizational function to the next.

Whether using the organizational or process mapping techniques, the first map should be an overview that illustrates the fundamental steps within the process. Refinement and details come in subsequent iterations.

After developing a comprehensive "as-is" process map of ACEG's returned goods process using these re-engineering hints, compare your map against the one developed by ACEG's re-engineering team as shown in Figure 4-3. How did you do?

*Where to Find Data*
Data for the mapping process should be gathered at the source from direct observation, interviews, process performance metrics, procedures, written policies, financial metrics, time studies, etc.

Metrics provide you with the ability to monitor a process and identify opportunities to improve both the cycle time and the quality of the process outputs. Data collection is the key. But data collection and analysis should not be burdensome. As with the process itself, always concentrate on keeping the data collection methods simple. The result will be less resistance from those responsible for collecting the data, and higher data integrity from those processing and analyzing the data.

*Assessing "As-is" Conditions for Improvement*
Now that the "as-is" conditions and the cycle-time model that represents the current operations have been developed, the next step for the manufacturing engineer is to establish a baseline cost

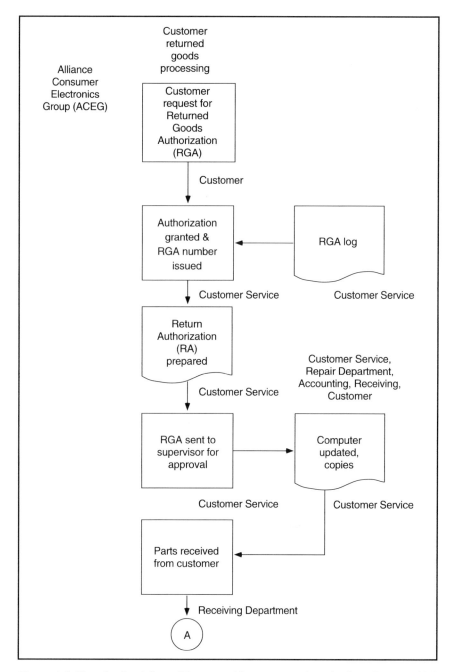

*Figure 4-3. Mapping "as-is" conditions provides a comprehensive baseline on which core process re-engineering personnel can structure a "should-be" map.*

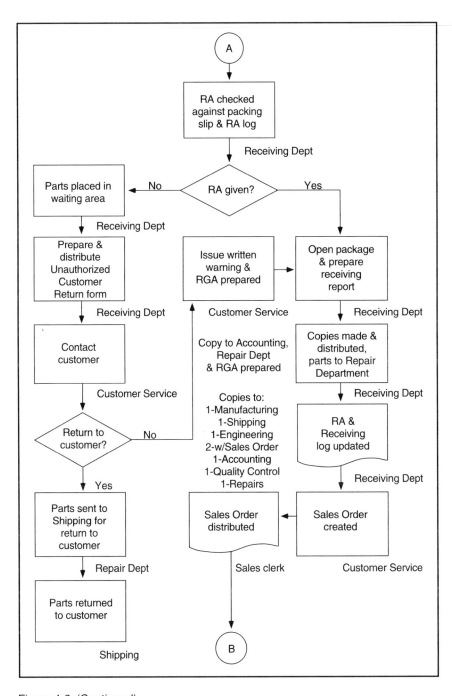

*Figure 4-3. (Continued)*

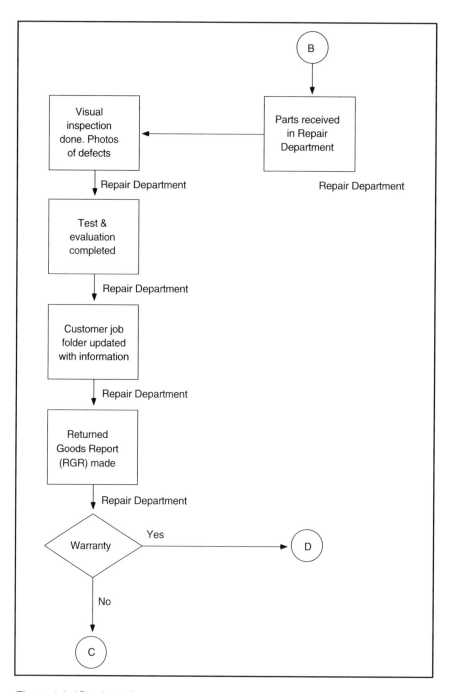

*Figure 4-3. (Continued)*

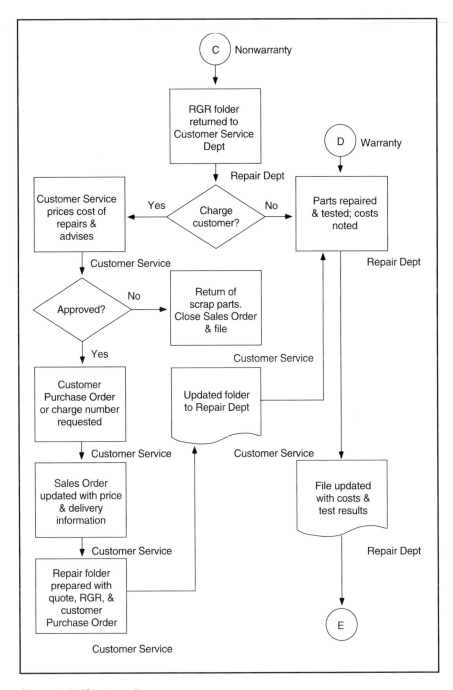

*Figure 4-3. (Continued)*

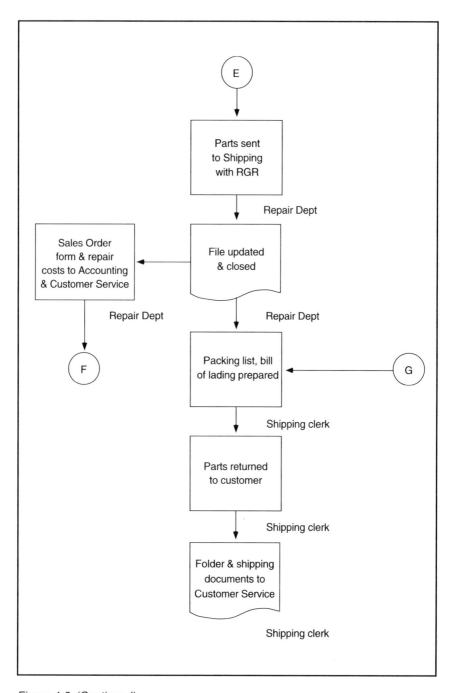

*Figure 4-3. (Continued)*

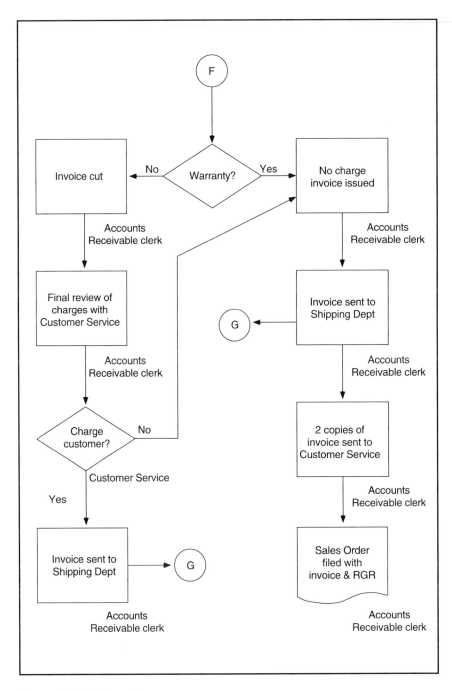

*Figure 4-3. (Continued)*

structure. Development of the "as-is" cost structure is important in that it provides the manufacturing engineer with a baseline from which to measure improvements. Remember, management wants to know certain things: "If I give you the money to re-engineer the processes, what will my return on investment be? How much will I save? How will those savings impact the bottom line?"

It's simple. For every investment, there must be an acceptable return. Otherwise, management will simply leave the money in the bank. The calculation of the "as-is" baseline costs, therefore, provides the starting point from which to measure how much—in tangible dollars—can be saved from each step in the re-engineering process.

A word of advice. When selecting the members for the re-engineering team, the manufacturing engineer would be wise to *select an accountant.* Here is where an accountant on the team can yield dividends. They know where the cost data is, they know how to calculate the savings, and they know how to develop the financial metrics needed to measure process improvements in financial terms. When it comes to numbers, they have the credibility. Capitalize on those benefits. They can help.

### Identifying the Hidden Costs

With the "as-is" map complete, it is a simple matter to accumulate the total cycle time of the process. This provides a steady-state picture of the cycle time of the process under optimum conditions. Here's the real test, however. Look for the "do loops," those steps in the process that say, "if this happens do this; if not, do this," or "if the information, product, or service is acceptable, do this; if the information, product, or service is not acceptable, do this." These loops are where cycle time—and costs—escalate dramatically.

Another hint: there is a tendency to discount time spent waiting for something in the process to happen; for example, the time the purchase order sits on the boss's desk awaiting approval. After all, you're not sitting there doing nothing while he takes his sweet time, are you? Of course not. But from the *customer's perspective,* time is passing and that time could be generating the customer additional revenues if he or she had your product in hand. Therefore, you must consider all elements of applicable time when calculating the total process cycle time.

One other thing. From the total cycle-time calculation, we're interested in determining the *total cost* of the process. That means we must consider all elements of value-added and nonvalue-added cost in the process. Items such as the following must be factored into the total cost equation for the process:

- Inspection costs: receiving, in-process, final;
- Test and confirmations costs;
- Design errors;
- Specification errors;
- Bill-of-material errors;
- Release delays;
- Overtime;
- Redesigns: product, tooling, process, packaging;
- Field failures;
- Field recalls;
- Warranty costs;
- Rework;
- Rejects;
- Time spent waiting or idle;
- Rescheduling of work, service, or production;
- Packaging errors or failures;
- Excess handling or processing time;
- Excess or obsolete inventory;
- Purchase order changes;
- Material receipt delays;
- Premium freight;
- Customer returns; and
- Excess use of energy and utilities.

Get the picture? These are only a few of the waste factors you will encounter as you begin analyzing the "as-is" process map. The rule: don't overlook anything. Question everything. Just because it's been done the same way for 20 years doesn't mean it's right for today or, especially, tomorrow!

*Assessing Control Effectiveness*
The assessment of the "as-is" conditions should include an evaluation of the controls, the metrics, and the reporting mechanisms used at the process and individual levels. The basis of the analysis is the internal and external customer matrices discussed earlier in

this chapter. The voice-of-the-customer analysis will define the requirements, expectations, and priorities of the process customers. From the VOC results, process metrics currently being deployed can be matched against those demanded by the customer; the gap between the two being the basis for the corrective actions required to take the process from the "as-is" to the "should-be" or targeted status. The Process Management Worksheet (Figure 4-2) provides the structure for this analysis.

As performance gaps are identified, the manufacturing engineer and his or her re-engineering team should conduct fundamental problem-solving analyses to isolate the root causes of the performance problems. Such analyses should include the inputs to the process to ensure the upstream process controls feeding essential value-added information, materials, and products are in place. In addition, the outputs of the process should be compared against the requirements of the internal process customers to ensure that all measurement systems are compatible (customer and supplier measuring the same things in the same ways), that all customer needs and expectations (explicit and implicit) have been identified and are being met, and that the process capabilities are in direct alignment with the requirements of the customer.

**Assessment hints.** Several hints are helpful during the assessment phase of the re-engineering process:

- A process output is evidence of completed work, not a description of the process itself or the activities contained therein, nor how well they are functioning. Tear the process apart to find these things out.
- Customers are found both inside and outside of an organization or work area. It is vital that they are all identified along with their respective requirements and expectations.
- People who perform the work generally know the process the best. Involve them in the validation of the "as-is" map and the creation of the process improvement plans.
- Customers have multiple expectations. Be sure to identify all of them.
- Identify the process improvement expectations to the process input suppliers and obtain their buy-in.

    *(Remember, the purpose of this phase is to identify opportunities for process improvement based on sound performance*

*data. Then, the most cost-effective opportunities can be addressed in the proper sequence, thereby maximizing the return on corporate resources.)*

- Monitor the process on an ongoing basis through formal measurements and observations. If any metrics exist, they should be validated.
- Compare process performance data to the expectations and requirements identified by the VOC analysis.
- Convert the data into information so that improvement opportunities are clearly defined.
  - Who are the customers and what do they want?
  - What are the known facts?
  - What additional information is needed?
  - What has been attempted before to improve the process?
  - Are the correct people involved?
- Brainstorm possible problem causes.
  - First, go for quantity.
  - Second, review the list for duplications, groupings, improbable causes, and clarification.
  - Third, prioritize possible causes based upon probability.
  - Fourth, verify each possible cause.
- Determine the economic value of capitalizing on the process improvement opportunities that have been identified and establish priorities based upon a Pareto analysis of them.
  - The first solution may be too costly.
  - The second may be too difficult to implement.
  - The third may be the best overall solution.
- For each solution, also define the nonfinancial issues impacting implementation.
  - Time required,
  - Resources required,
  - Political repercussions,
  - Risk.

Now, establish the targets of improvement based on a critical assessment of the potential process capabilities after implementation of the identified corrective actions, coupled with the requirements established by the customer in the VOC analysis. From these targets, identify performance metrics for the process that must be re-engineered into the new process to ensure continuous conformance to customer requirements. A valid starting point is the actual metrics the customer will use to assess the process outputs (performance, perception, and results).

### Re-engineering the Process: The "Should-be" Map

At this point, the assessment of the "as-is" conditions has been completed including identification of the root causes of the process problems. It is, therefore, time to re-engineer the process into one that is more time- and resource-sensitive. Here are a few hints to facilitate the re-engineering efforts.

- Take a clean-sheet approach. Too often, the re-engineering team will focus its efforts on optimizing the existing process versus taking an entirely fresh approach. The dangers in that methodology are that the team will take for granted that each process step is required, in the same sequence, and governed by the same process owner(s). Its focus will simply be to change the metrics, resulting in the same old process with new guidelines. There may be some incremental improvement, but no significant gains will result from this approach. *Take a clean-sheet approach.*

- Question everything. Ask "why" at least five times for each step in the process. Do not accept the "it's worked fine for years" response. Listen to the suggestions from those members of the re-engineering team who are least familiar with the process. Their input will likely lead to thinking about the process in a new way. Benchmarking is also an excellent tool at this stage in the re-engineering process. Find out who the best-in-class companies are for the type of process your team is addressing and tailor their methodologies to your needs.

- Simplify the process as much as possible. Simplicity enhances both the time and workability. Wherever possible, eliminate process steps or combine multiple steps into one. Watch for those elements in the process that involve queue time, wait time, setup time, and move time. Watch for information or material batching points, inspection/review points, approval points, etc. Be mindful of distance between process steps: the longer the distance, the more time it will take.

- Every step in the process is fair game, there are no sacred cows. If you meet resistance, it is a sure sign that process step needs further analysis.

- Focus on innovation, speed, quality, and customer service in every step of the process using the VOC results as your guide.
- Old assumptions about the business, the process, and the customers are just that, old! Create a new order. Times have changed; technology has changed; customer performance expectations have changed.
- As you re-engineer the process, build capability and control into it. Be sure all decision points are at the lowest level possible, focusing on the individual level as often as possible. To be effective, decisions should always be made at the level the work is being done.
- As you develop the process, consider what skill sets will be required to maintain the new process and what training will be necessary to support those skill sets.
- Don't forget the value-added/nonvalue-added reviews as we defined them earlier. If an activity, operation, or entire process does not add value to the customer or the process of satisfying the customer, eliminate it.
- Remember to base your re-engineering proposal on facts. Data is essential, but it must be good data. Always validate your information through first-hand observation, measurements, data bases, interviews, etc.
- Consider running process steps concurrently versus serially. It maximizes the amount of value-added activity that takes place within the same time span.
- Where decentralization is involved, and cannot be designed out of the process, utilize electronic methods to speed the communication and information flow . . . electronic data interchange, bar coding, groupware, teleconferencing, facsimile, etc.
- Eliminate as many "approval" processes as you can. The internal controls you have built into the process should handle most, if not all, requirements for review and approval.

*Documenting the New Process*

As the re-engineering team finalizes the new process, it is imperative that the process documentation be completed. The documentation should include process

- Metrics,
- Scope and capability,
- Steps/flow,

- Dimensions,
- Owners,
- Objectives,
- Controls, and
- Timing and cost.

Process- and individual-level procedures also should be included. In addition, at this point it is time to determine the savings in time and dollars the new process provides over the old. This is important to getting senior management's approval to move forward with the implementation. If we had not determined the baseline of the old process earlier, we would have had no way of completing this vital step, and the team's work may have gone for nothing.

### A Time for Introspection

Is our approach reasonable? Is it workable? Is it implementable?

These are important questions. The manufacturing engineer and the team should seek an outside opinion (often from the customers and suppliers to the process) at this point relative to the validity and sustainability of the re-engineered process. A key issue is whether the new process has a positive or negative impact on both upstream and downstream operations. The solution is not valid if it optimizes one process at the expense of another; or worse yet, compromises performance at the organizational level.

Seek critical assessments of the team's work. If there is a problem, it is better for the team to find it *before* going to management for approval to proceed. Once the re-engineering team is comfortable with the outside assessments, and any corrective actions resulting from the assessments, then it is time to proceed with the development of the implementation process.

### Implementation Planning

Certain components are critical to development of all new-process implementation plans.

- Identification of who the owners of the implementation process will be. Because there must be both responsibility and accountability for results, the process owners and the re-engineering team must be actively involved, along with any internal suppliers, internal and external customers, and process mentors or sponsors.

- The implementation plan must identify from whom approval must be obtained in order to initiate the implementation process. Political roadblocks can be disastrous to both the re-engineering process and the careers of those involved. To obtain approval, the team must identify the "benefits" of the re-engineering process to each of the individuals from whom approval is required. Then it's up to the team and its mentors to sell the process throughout the organization.
- Often, the re-engineering process will dictate a change in the structure of the organization itself. In such cases, the re-engineering team must identify the resources impacted, any associated reallocation of those resources, reallocation of responsibilities and associated impact on that function's or individual's workload, any changes in job descriptions or classifications, any location changes required, and all training needs associated with the process changes.
- A master timetable, which defines the events and activities which are to take place, and the timing of each, is an integral part of the implementation planning. It should be comprehensive and based on a concurrent versus serial implementation methodology to compress the change process into the shortest possible time. Why? Change is often painful and it consumes valuable resources. By compressing the time frame, the team can minimize the  pain of change and its impact on valuable resources.
- Metrics to monitor the implementation and measure its results are another vital element in the implementation planning. The metrics should be based on the savings commitments made by the manufacturing engineer or the team along with the VOC requirements defined earlier in the re-engineering efforts.

Remember, change is an unwelcome event to most people because it forces them out of their comfort zone and into an arena in which they are not familiar. There will be resistance. The best way to combat that resistance is through a comprehensive implementation plan that is executed flawlessly.

*Implementing Change*
When given the opportunity to change or continue with the current unpleasant circumstances, most people will begin to rationalize why their current circumstances aren't so bad after all.

From our earlier assessment of the resistance to change in your organization, we have identified the degree of resistance and the associated level of risk in implementing this type of change process. From this, we can also identify the pockets where that resistance exists. If the manufacturing engineer and his or her re-engineering team have done their jobs effectively, these factors have been taken into consideration in the development of the implementation plan and the associated training requirements contained therein.

As with all sales (and this is a sales process), timing is critical. We have identified from whom the team must obtain approval and the triggers to ensure their buy-in. Now it is a matter of timing and support from the team's mentors. Working through the team's mentors, schedule a convenient time to meet with each individual from whom approval must be obtained. In advance of the scheduled meeting, forward a synopsis of the team's work, recommendations, benefits, and savings for that person to review. At the meeting (the team's mentor should be present), the team should provide a concise overview of the implementation plan and its costs, timing, metrics, and expected results. Any supporting documentation should be available in the event it is required to support the team's findings or recommendations. The presentation should be structured to answer any questions which could arise so as to disarm any objections the individual may have. The team should request that person's approval and support upon conclusion of its presentation.

Resistance can be both politically and personally motivated. If your team meets with either, try to assess the basis of the person's resistance. Then restructure the implementation approach to alleviate that person's concerns. Use the team's mentor to push his or her peers into acceptance of the plan. When all else fails, try a top-down approach. People find it difficult to say no when the big boss says yes.

Implementation should follow the agreed-to plan. Any deviations should be approved by all affected parties in advance. Implementation metrics should constantly be monitored by the team to ensure they are on target and remain focused. And when it is completed, it is time to start again. Change is a never-ending process.

*Let's See How You Did*

Compare your "should-be" map with that created by ACEG's re-engineering team, Figure 4-4. Were your ideas, and your ultimate solution, as effective? As creative? (Again, consider it from the *customer's perspective*.)

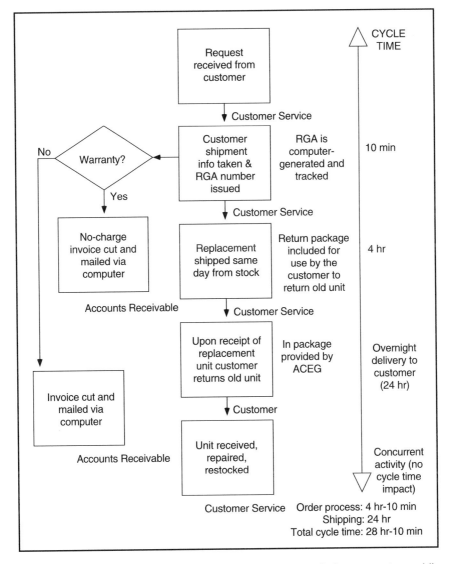

Figure 4-4. An effective "should-be" map eliminates unneeded process steps while retaining or improving efficiency levels.

# 5

# Fundamentals of Quality Design

Quality comes from designs that incorporate *producibility, maintainability,* and *recyclability* into the process. No design can guarantee a high-quality, cost-effective product unless the product can be easily produced by the manufacturer and is of value to the customer.

Ever wonder why design engineers seem to constantly design products that cannot be made economically using existing tooling and equipment? The answer is more obvious than you might think: no one has told them what capabilities to design for. Ask yourself these questions:

- Are your design engineers aware of the capabilities—and limitations—of your internal manufacturing, fabrication, and assembly operations from statistically-based process capability studies?
- Are your design engineers aware of your suppliers' process capabilities from a similar statistically-based process capability study?
- Does your company utilize an approved supply base listing which restricts specifying products from other than those suppliers included therein?
- Has your management established a strategic operations plan that defines both the products and manufacturing processes required to support the strategic 5-year business plan?
- Have design rules been established jointly between design, quality, and manufacturing engineers that incorporate sound design-for-producibility and -assembly principles structured around the process capabilities available to you and consistent with your organization's strategic direction?

If any answer is "no," you cannot assign your manufacturing problems to Design Engineering. Nor can you blame them for overly cautious or "safe" designs. Given none of the above information, and a constantly moving customer specification target from Marketing, any design engineer will structure his or her designs with a high degree of robustness and tolerance safety. It's only natural. You and I would do the same.

So how do we break the cycle? We help the design engineers understand what works and what doesn't from a manufacturing viewpoint. Our objective is to maximize through the product design our capability to produce the product economically, considering the most efficient methods of fabrication, assembly, test, inspection, installation, and maintenance. To do so, each functional requirement of a component or subassembly must be met effectively by some aspect of the design itself. Industry literature is replete with design rules and guidelines that are now practiced by many world-class organizations. These companies use the guidelines to ensure that their designs are both economically producible and conform to customer requirements and expectations. We will review those design rules shortly.

In addition, we must also recognize that in the future only product designs that are environmentally benign will be acceptable to the customers. The beginning of that trend has already been seen in the ISO 9000, ISO 14000, and QS 9000 standards which have effectively shifted the burden of proof for product safety to the manufacturer from the consumer. In addition, most product designs are now required by the customer to incorporate upgradability, easy disassembly, and recyclability. Take for example the automotive industry. By the 1994 model year, many U.S. and foreign automobiles had reached the level of 85%-plus recyclability, and that trend is expected to continue. Think what that means for the general population as we begin to reduce the need for landfills and consumption of our vital natural resources. Think, too, about yourself as a consumer. No more buying of an expensive computer or electronic device only to have it become obsolete within months after the purchase. If you really want to have fun with this, think about what design-for-assembly means to you as a parent! Remember last Christmas Eve? Did you spend it as I did, trying to assemble toys that were designed such that "any adult with only simple household tools can assemble them"? Right! And we're engineers.

The requirement for agility will dominate many future designs, forcing the requirement for customization under user-defined criteria, in quantities as few as one, at cost/benefit ratios established by the customer versus the manufacturer. As we near the 21st century, more and more products will be configured to the customers' requirements at the time of sale. Consequently, the manufacturing and design engineers must position their organizations to be capable of responding to those challenges. The time to begin is now.

## QUALITY DESIGN REQUIREMENTS

Sound design rules always focus on simplicity in both the design and manufacturing processes by eliminating components and process steps whenever and wherever possible. The objective of the design is to create a product that meets customer requirements and expectations, while reducing tooling and processing costs, thereby improving quality, reliability, life-cycle costs, and new product development cycle time. Such designs thus consider everything from the planes and methods of insertion to methods of joining and fixturing, and from assembly rotations to tool design and tool life. In short, every element in the new product development process from design concept through production and installation of the product, including post-installation maintenance, is included in the design function.

The following lists some of common-sense design criteria derived from numerous sources within the manufacturing community. If applied to your products and processes, the quality and cost of your products will improve significantly.

- Clearly and completely define the design requirements with the customer and the concurrent engineering team before beginning the design work. With fewer downstream changes, higher quality of both the design and the product will result, as well as lower cost, and shorter time to market.
- Make sure the design incorporates the quality inspection criteria and processes for the components and subassemblies. Don't rely on quality engineers to come up with final inspection methods to ensure zero-defect levels after the product has been released to production. It simply won't happen. This means that the final drawings must identify everything from critical

characteristics to the frequency of sampling those characteristics in line with proven statistical sampling criteria.
• Incorporate design of experiments (DOE) and Failure Mode
  and Effects Analysis (FMEA) methodologies to optimize the
  designs and corresponding production and quality processes,
  with the objective of "designing out potential quality problems
  before they can occur." Prevention is always less expensive than
  failure and appraisal.

Table 5-1 illustrates the escalating nature of the cost of a
nonconformance. Given your particular industry and business,
you can factor the figures up or down accordingly, but the ratios will remain constant.

### Table 5-1. The Cost of Nonconformance

| The Cost to Resolve a Design Defect | |
| --- | --- |
| In design is | $    250 |
| In production planning is | $  2,500 |
| In manufacturing is | $ 25,000 |
| By the customer is | $250,000 |

The numbers speak for themselves. You, as manufacturing
engineers, know the devastating impact of problems and design defects on the manufacturing processes, on tooling
changes, on your time. The key to resolving those issues is
prevention.  I have used value analysis (VA) for years and respect what it can do for cost reduction. But I also realize VA
techniques have one flaw: *They are after the fact. The problem has already occurred.* Fully 90-95% of the cost of a product is designed in. Just think what happens every time you
redesign it.
• Minimize components. The fewer the parts and subassemblies,
  the lower the manufacturing resources required to produce
  the product, plan production, purchase and handle the raw
  materials, and the lower the inventory, tooling, and fixture
  costs, etc.
• Maximize the functionality within each component and assembly by combining the functions performed by several parts into
  one whenever possible. True, that may make that part more

complex to tool, but the total tooling and inventory costs may well be substantially less. Always consider the *big picture*.

- Wherever possible, use gravity as an assembly aid by designing parts that can be assembled in a single top-down direction. Such practices allow you to either eliminate or simplify the holding and fixturing requirements because the base components often can be used as the fixturing device. Top-down design techniques also facilitate automation and robotics. But remember, automation capabilities must be designed in, not added after the fact.

- In addition to the top-down methods, consider insertion requirements and methods. Always try to avoid components that require multiple insertion operations or methods. It complicates the assembler's job and increases both cost and quality problems.

- Always design products to minimize the number of surfaces that must be processed. This facilitates component location and placement, while minimizing wasted motions and time to reorient the component several times during the assembly process. Quality is improved, as is the ability to automate the processes; as a result, the cost of automating the processes is reduced.

- Don't design yourself (and the assembler) into a box. Minimize restricted or blind access operations. Always provide the fabricator or assembler with a clear view of his or her work. "Assembly-by-feel" designs often lead to poor quality, excessive cycle times, and safety-related costs. To illustrate the point, take a trip to your local foundry or injection molding operation. Watch the cleaning operations. Compare the time and effort required to clean the external surfaces versus the hidden surfaces. Or look under the hood of your car for access to the spark plugs and oil filter.

For years, I have enjoyed restoring classic muscle cars. You may remember the mid-60s to late-60s Mustangs with the big-block high-performance engines. Like me, I'll wager you have probably forgotten the maintainability problems that went along with those "toys." Recently, I began a restoration project on a classic fastback. After a frame-up restoration and reassembly, I began the final finishing touches on the engine compartment—the carburetion, points, plugs, filters, voltage regulator, wiring, etc. One of the final steps, was to replace the old plugs with newer-generation platinum plugs that would operate well given the performance demands of the engine and the current octane ratings of available gasolines. So, I pulled out the old plug socket and wrench, then headed to the car for what I had anticipated as a 30-45-minute job. Surprise! To access the plugs, guess what two options I had? Option one: drill holes in my new inner fenders for an 18-inch extension, or, two: pull the engine out so that I could access the plugs with conventional tools *and not destroy all the work I had done to restore the car to its original condition.* The point speaks for itself.

• Avoid designs that incorporate numerous fasteners, or a number of types and configurations of fasteners. Fasteners are a major contributor to extended assembly cycle times and maintainability problems. In fact, loose fasteners are the single largest contributor to equipment failure, even over lack of lubrication. So why design-in problems? In addition, fasteners are difficult to feed automatically, are often intermixed during line stocking operations—causing their misuse and bill-of-material related stock-out problems—and require 6 to 10 times their cost to tighten. Every configuration requires a separate tool, thus extending changeover times and making the assembly operations more complex. In many instances, the use of interlocking or snap-fit designs are much more cost effective when the total cost is considered.

   Recently, a gang of veterans at one of the Big Three automakers did an excellent job reducing the number of fastener types on a new car design. The new model has about 300 kinds of screws, bolts, and other threaded fasteners, versus 650 on an average car. A significant accomplishment, yes, but there still remains considerable room for improvement.

• Design for unique part identity. Having once been a material handling manager, I readily identify with this one. Parts that look similar are often misidentified, incorrectly stocked, and incorrectly used, resulting in numerous shortages that create increased expediting and material handling costs. What is worse are individual parts that have multiple part numbers. Those can never be kept straight.

- Simplify part orientation. Symmetrical parts often are designed to facilitate automation and orientation during assembly. The simpler the orientation, the easier to assemble, the lower the cost, and the fewer the quality problems. There are times, however, when part identity and orientation are critical. In those cases, the use of asymmetry on exterior and mating surfaces is normally recommended. It enhances both the visual recognition of the part, and the associated assembly operations.

- Remember, always design for automation versus attempting to automate an assembly operation once the design has been completed. At an agricultural equipment manufacturer I worked for some years ago, I recall the first robot welder we purchased. It was acquired to reduce the welding time on steel plates to overcome both a capacity and a quality problem (in order to alleviate the quality defects and weld breaks, we had previously increased the time standards so the welders would take more time to ensure they were producing good welds). The robot was easily justifiable with the projected reduction in standards and rework costs, so we placed the order, cleared an available area in the fab shop, painted the floors, erected the robot, built a safety fence around it, and went to work. A year later, there it stood looking as good as the day we installed it. Why? Because we never used it. We failed to follow one (at least one) of these cardinal rules. The robot failed to weld straight, sound welds consistently. The blame was placed on the robot: unproven technology; poor equipment design; too complicated; etc. The fact was, the parts we were trying to weld were inconsistent. One time they were perfectly square per print, the next time they were out of tolerance. The robot didn't know the difference, so it tried to weld each piece the same way. (Can you imagine that?)

- Design for assembly. Effective designs focus on the assembler and the assembly processes. Those that incorporate lead chamfers, for example, aid in overcoming misalignment problems while facilitating automation. Always consider what people do well and what machines do well when designing a component or assembly. Remember, many assembly operations are done best by people, while pick-and-place operations are often better accomplished by robotics.

• Avoid designing material handling and assembly nightmares. Springs with open loops or ends will tangle, as will compression springs. Lock washers fall into the same category, as do parts that interconnect or nest. Here too, think about the assembler. Watch what an assembler goes through to untangle a handful of springs or other interlocking parts when he or she pulls them from the parts bin. Not really time effective.

• Always design to readily available process capabilities: those that exist in the processes you currently use in your manufacturing operations and those that are available through approved suppliers and subcontractors.

• Whenever practical, standardize components and processes across product families; it not only facilitates group technology for reduced setup and changeover times, but provides the capability to employ modular bills of materials, tooling, fixtures, and subassemblies that, in turn, reduces inventory investment and handling costs. The use of modularized bills (sometimes referred to as planning bills) facilitates the planning and procurement processes, as well as the engineering change process.

• Avoid designs with adjustments, especially those with multiple adjustments. Adjustments complicate both assembly operations and field use by the customer. They require detailed methods and assembly procedures, increased sophistication in quality assurance, and expanded test procedures and processes.

Remember the old television sets? The ones with all the knobs on the front so that you could adjust everything from the color tint to the vertical hold, from the contrast to the horizontal hold, all to make the picture just the way you wanted it? What did you do with those knobs? You turned them, right? Then what happened? You could never get the picture back the way it was in the beginning. So you complained. And the manufacturer, weary of hearing your complaints, moved the adjustment knobs first to the back of the set where you could not reach them (but you found them anyway and messed with them), then behind concealed compartments (but you found them, too, and messed with them). In total frustration, the television manufacturers eliminated the adjustment knobs altogether and gave you *remote controls* so "they" could control

what you messed with. And you thought it was a product enhancement so that you would not have to get off your duff to
change the channels, right?

- Always design for ease of handling and packaging (from the
consumer's perspective). We're engineers, right? We like gadgets. We buy gadgets, lots of gadgets. And because we're good
engineers, we always assemble everything we buy ourselves.
Sometimes we read the instructions before starting, sometimes
after we're through (to find out what these extra parts are for).
In either case, we ultimately rely on the instructions at some
point. That's where this rule comes into play.
- The latest designs for small products incorporate a shrink wrap-
ping process in which the product is placed against a multi-
purpose card which is used to hang the product for display, to
illustrate the assembly instructions, and to prevent pilferage
and lost parts. The product(s) and the backing are then wrapped
with a plastic overlay and heated to shrink the overlay, thus
holding everything securely in place. The problem? To guarantee it all works as designed, the package engineer specifies
a robust 150 mil, bullet-proof cut-proof vinyl. A little more expensive, but it ensures no shop lifters will get away with their
products! But alas, along comes the consumer and buys the
robust package to get the goodies inside. If they are like I am,
by the time they get the darn thing apart, they have completely
destroyed the instruction card.

  The key here is to remember the intended use of the package, and to design it with just enough robustness to do the job
without creating additional problems for the consumer.
- Empathize with the assembler. Designs that are self-nesting
and orienting are simpler and faster to assemble. Think of how
the pieces of a puzzle snap together. They only go together one
way; the stress planes are addressed effectively; they require
no secondary processes like adhesives, fastening, stapling,
welding, etc.; the components are easily identified and oriented;
the assembly process is unidirectional, typically utilizing gravity as an aid; and they are self-fixturing. Simple, yes. So why
not use the same design approach whenever feasible in your
operations?
- Don't forget the maintainability aspects of your product design. If the product requires disassembly for normal mainte-

nance procedures, then make disassembly an integral part of the design. If the product requires periodic lubrication, then make the lubrication points accessible. Do not force the customer to disassemble more than the absolute minimum number of parts to complete the required maintenance operations, using readily available tools. If it is complex or requires special tooling, either the maintenance will not get done, or the customer will fail to do it correctly. Either way, you lose. Remember, not everyone who likes to do their own maintenance is an engineer.

One of my client organizations is a prestigious university in the South. The architecture of all new facilities on the campus is closely controlled to ensure continuity with the southern theme of the original architects and designers. Recently, a new building was designed and constructed. When the design was originally conceived, the HVAC units were located on the roof. The university's building committee rejected the design as inconsistent with the architectural theme of the university. The architects then suggested the HVAC units be placed in the basement of the building with only the condensers placed external to the building. The committee accepted the architect's recommendation with the caveat that the condensers be placed on the rear of the building rather than the roof so as not to be visible to passers-by. And so it came to pass. And whenever the facilities crew were required to clean the coils or change the filters, they rented two mobile cranes. One to hoist the maintenance men up the side of the building where they could access the units, and the second to hold up the outer shell of the condensing unit while they completed their maintenance activities. An operation that would normally have taken two to three hours to complete, now takes seven to eight, at six times the normal cost (but the building looks good!).

Here are a couple of facts for you to consider when designing your products.

1. Fully 95% or more of your unhappy customers will never let you know the reason for their displeasure. They will, however, tell at least nine other people. Thirteen percent of them will tell 20 or more, and

2. More than 90% of your unhappy customers will never buy from you again. They will simply cease to exist as far as you are concerned.

Sounds more and more like we have one shot at the target. A single failure can be catastrophic.

## THE FINAL NPD SYSTEM ASSESSMENT

The role of the manufacturing engineer doesn't stop once the design work is completed, manufacturing processes are finalized, tooling is planned, and standard costs are calculated. Before we can release the design to production, the manufacturing engineer must complete a final new product development (NPD) system assessment (Figure 5-1). Why? To ensure that the final design has been accurately documented, that all design rules have been adhered to, and that all enabling support systems are in place.

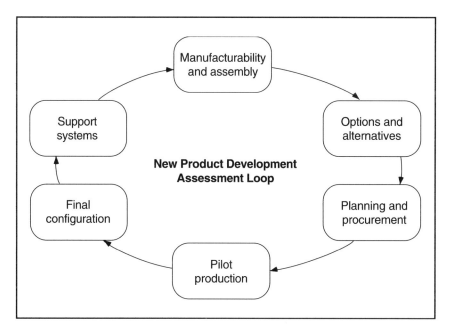

Figure 5-1. The manufacturing engineer's involvement in the process pervades the entire manufacturing cycle.

The manufacturability and assembly assessment includes a comprehensive comparison of the alignment between the product design requirements and internal production capabilities to ensure that the product, as designed, can be built within quality, cost, and lead-time targets (Figure 5-2). Also assessed are the capabilities of the supply base to ensure that the required raw materials and components can be supplied consistently within quality, quantity, cost, and lead-time parameters.

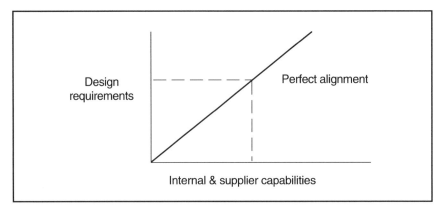

*Figure 5-2. Final new product development system assessments will reveal how well production capabilities match with design requirements.*

Any nonconformances or questions must be addressed before moving forward. If left to be addressed later during the production cycle, significant performance and cost risks will be introduced to the manufacturing processes. Also to be assessed at this stage is the alignment of the organization's technological competencies against those required by the design itself. Can the product be produced using readily available equipment, tooling, and processes? Are tolerances and specifications attainable? Have the results of FMEA and DOE analyses been incorporated into the design and supporting production processes? Have tool and fixture designs been validated, along with test and inspection fixtures and methods?

Financial alignment should be reviewed again at this juncture to confirm that the product can be produced and distributed within the price-point targets established by the marketplace, and still provide the organization with an acceptable margin. And finally, a critical assessment of the design's conformance to customer requirements, priorities, and expectations must be completed before moving into the next stage. If differences or failures exist, now is the time to reconcile them—*not* after the product is in production, or worse yet, in the hands of the customer.

### Options and Alternatives Assessment

In a perfect world, option and alternative assessment would not be necessary. Unfortunately, mistakes are made, customers make last-minute changes, and the best of designs occasionally fail to conform to performance criteria. Therefore, it is wise to have options available in case the product or one of its subassemblies proves unproducible or too costly. Obviously, the earlier in the NPD cycle this condition is identified, the less an impact the ensuing design changes will have on both the cost of the design and the ability to complete the NPD cycle in accordance with planned schedules.

On highly technical or complex designs, consideration should be given to running parallel design processes. Admittedly, there is a cost in so doing, and it must naturally be a factor in your decision, but so must the tradeoffs, should the original design fail to live up to expectations.

### Planning and Procurement Assessment

The comparison of internal and supply-base capacities against forecasted market demand for the new product is critical at this stage in the NPD cycle, in terms of allowing for the up-front identification, allocation, and deployment of the resources required for both the production ramp-up, and the distribution end of the cycle. The checklist of considerations should include:

- The master production plan
  - Can the forecasted volumes and product mixes be produced using only available or planned capabilities and capacities, or must additional capacity be allocated or obtained ?
  - Have the pilot and production release dates been confirmed, and are they attainable?
  - Has the availability of raw materials been confirmed and is it consistent with the pilot and production schedules?
  - Has the availability of required tooling, fixtures, and equipment been verified, and does that confirmation include erection, test, calibration, and validation?
  - Have service parts and post-sale service requirements been considered and adequately addressed?

- Has employee training been conducted or scheduled on all new processes and equipment *prior to* the forecasted pilot production launch?
- All supporting documentation
  - Have the design rules been followed, with exceptions (if any) documented?
  - Have manufacturing routings, methods, policies, and procedures been developed or scheduled for completion prior to pilot production release?
  - Have test, inspection, and final confirmation methods and procedures been completed and validated?
  - Have the component and assembly drawings and bills been audited, with any resulting corrections made?
  - Has any required installation or maintenance documentation been finalized, and will it be available to meet forecasted product ship dates?
  - Has all safety, environmental impact, and warranty registration documentation been completed and printed?
  - Have engineered cost standards been developed and confirmed?
    - Are they attainable under "normal" production conditions?
  - Has all price book, sales, and marketing literature been printed and prepared for delivery to the appropriate distribution channels in advance of product introduction?
- Quality system confirmations
  - Have all critical characteristics been identified and noted on the product drawings, along with the frequency of inspection of each?
  - Have process control points been established and documented on the routers and methods sheets?
  - Have the quality assurance methods and procedures been confirmed, and have they been communicated to appropriate personnel along with any required training to ensure compliance with them?
  - Has all required inspection and test equipment been confirmed and calibrated to national standards?
  - Have all applicable certifications and approvals been received by the appropriate governing bodies and agencies, such as UL, CSA, ISO, EPA, etc.?

- Process cycle-time confirmations
  - Has the order entry process cycle time been measured and have appropriate metrics been developed to ensure continuous improvement?
  - Have all administrative cycle times been measured and included in the total quoted lead time?
  - Have production planning and material procurement cycle times been measured and included in the system lead-time calculations for purchased raw materials and components?
  - Have the manufacturing cycle times been validated and are they attainable on a consistent basis?
  - Have transportation, packaging, and handling cycle times been measured and incorporated into the quoted lead time?
- Distribution methods
  - Have field stocking or floor-planning methodologies been considered, along with their cost impact?
  - Have unit and bulk packaging methods been finalized, and the availability of any required raw materials, tooling, or equipment confirmed?
  - Has inbound and outbound transportation planning been finalized, along with the cost of shipping, including any applicable duties, tariffs, taxes, royalties, etc.?
  - Has the transfer or freight-on-board (FOB) point been established and confirmed?
  - Have all customer labeling or bar coding requirements been adequately addressed?

Overwhelmed? Hold on, there's more.

### Pilot Production Assessment

Once the planning has been completed and the design finalized, a pilot production run should be scheduled to ensure that all technological, resource, and financial issues have been resolved. In addition, the pilot production process provides a final check on the production readiness, tooling and fixture designs, raw material specifications, and quality system effectiveness.

In addition to internal considerations, the pilot run provides an excellent opportunity to confirm distribution methods, test packaging and labeling systems, test-market the product to confirm projected demand, and perform final on-site application testing of "production" products in the customers' environments.

### Final Configuration Assessment

Working with literally hundreds of organizations around the globe, I have found a truth that has become practically universal: the final product configuration rarely corresponds 100% to the engineering documentation. Normally in the rush to get the product manufactured and shipped, this final dotting of the "i's" and crossing of the "t's" is left to be done at a more convenient time. But that time rarely comes—at least not quickly.

The result of releasing products that do not meet the bills of material is not usually visible at first. It manifests itself as a series of ill-timed surprises that begin to emerge downstream. Surprises like:

- Unexplained variances in standard costs,
- Increased manufacturing cycle times due to unplanned secondary operations,
- Inventory shortages or overages resulting from imbalances in usage and demand, and
- Quality system nonconformances when products are checked against incorrect documentation.

And should you decide to purchase the product, or any of its lower-level components from an outside source rather than to make them in house, the headaches are compounded. Your suppliers will be making products to your specifications only to have you reject them, which makes for some interesting discussions if you thrive on conflict and discomfort. Or, their costs and your costs will never coincide, so you will never be able to justify outsourcing regardless of your capacity problems.

To prevent these from becoming issues, consider a final configuration assessment. This process should verify that:

- All bills of materials, product specifications, process and tooling designs, routers, fabrication and assembly instructions, quality procedures, special handling instructions, and product costs are complete, accurate, and consistent with the configuration of the product itself and the processes used to make it;
- All price books, sales brochures, and field installation and owners manuals have been printed and accurately reflect the final product configuration and its available options;
- All product application and compatibility issues have been researched and resolved prior to product release;

- All product packaging has been reviewed and meets the requirements associated with product distribution, point-of-sale display, and product protection.

### Support Systems Assessment

When all product system assessments have been made, everything should be ready—the documentation, the product, the production processes, all producibility issues, distribution and packaging methods, quality assurance processes, materials and tooling, etc. Prior to final release, the manufacturing engineer and his or her NPD team should ensure that all required support systems are in place to guarantee a smooth transition from pilot to full production.

This final step requires a review of all quality systems, manufacturing systems, planning and scheduling systems, procurement systems, personnel and training systems, distribution systems, aftermarket support and field service systems, and customer service systems. The purpose is to determine whether the organization can fully support the launch of these products now and for the foreseeable future at the projected volumes and product mixes, without stumbling. If the answer is yes, then it's time to start your engines.

## CLOSING THOUGHT

In the next chapter, we will explore some of the methods currently in use to reduce product setup and changeover times. You have probably noticed that much of the emphasis in this book has been on reducing the *time* elements in our businesses. That is exactly correct. Why? Because time is a valuable and costly commodity, one that cannot be squandered or arbitrarily allocated to activities that do not generate an adequate return on investment. Your responsibility as a manufacturing engineer is to focus every activity (administrative and operational) on operating effectively, because every activity in your operation directly or indirectly contributes to lead time, quality, and customer satisfaction.

# 6

# Optimizing Lead Time and Productivity Through Effective Setup Reduction

## TIME: FRIEND OR FOE?

As we discussed in the opening chapter, in many cases today, quality is not the single largest factor in the buying decision. That's because today, quality is *expected*. Without it, you are not even considered by the consumer. Today, an organization's quoted (and actual) lead time has become the largest differentiator between providers of products and services. And because of that, manufacturing engineers must focus on re-engineering the core changeover and setup processes to allow their organizations to compete through reduced lead times and increased quality levels—all at lower internal costs.

The key to effective lead-time management is the elimination of all elements of waste in operations. Waste drives up costs, increases manufacturing cycle times, and reduces quality. Waste infiltrates every facet of your business, every function, every process. Waste has been identified as the common denominator in business failures within a broad range of industries.

### What Exactly is Waste?

Waste is the excessive or inappropriate consumption of corporate resources—employees, machinery, materials, and capital—beyond what is absolutely necessary to produce products. Examples of waste in the manufacturing processes typically include:

- *Surplus inventory*. How much is enough? How much is waste? Let's put it this way. If you don't have more than 24 to 30 inventory turns annually, you have too much inventory. Many manufacturers today, even the smokestack metal-benders, have more than 50 inventory turns a year. So where does this put your company? Remember, inventory is an investment that yields returns only after it is converted into sellable products. Until then, it is a corporate asset with a negative return on investment.

- *Producing more than is required to meet customer demand.* "Well, it was on the machine anyway." "May as well run it out." "We'll need them in a couple of months." "Got to keep the machines running 24 hours a day to ensure we get the best payback on our investment." "It's critical that we generate those direct labor hours for absorption." You can fill a book with the reasons for overproducing. (Funny thing is, most companies sell "stuff," not direct labor.) Ever wonder where those direct labor hours go? Right into inventory. So the more excess labor that is generated, the higher the inventory. It's a real fallacy that most standard cost accounting systems perpetuate. And until it is addressed through an activity-based costing system, or similar actual-cost system, it will continue to work against your organization.

Then there's the old standby: "We must run as many parts as possible on this setup because setups take a long time. Got to generate more direct labor to get the highest absorption possible so the budget looks good this month. Inventory? That's not my problem. We're interested in labor productivity, not inventory."

Let me give you an example of a capital goods manufacturer that followed that approach. The company's productivity was poor (actually it was horrible), so it followed Accounting's advice and doubled lot sizes to minimize the time spent on setups. Two things happened shortly thereafter. One, inventory (raw and work in process [WIP]) increased by more than 40%. Two, productivity still didn't increase appreciably because the bottlenecks in the shop just became more backlogged (nothing had been done to fix the bottlenecks, so they simply became even greater problems). Know what else happened? Business dropped because lead times were much too long and delivery performance was extremely poor.

Here's a little tip. When setup times are effectively managed, lot sizes can be reduced. When that occurs, shop throughput increases, driving up productivity per direct labor hour. Cycle times decrease, thus reducing customer order lead times. Quality increases because there is a natural tendency to place more attention on the quality of one's work when there are only a few pieces in the lot versus many (the theory behind JIT), and operating costs decrease. And, *inventory levels decrease*, driving up inventory turns. A true win-win scenario.

• *Expending energy and resources making bad parts.* "Well, they are just setup pieces anyway. All that scrap won't go to waste, we'll use them for setup pieces. Don't worry about getting the setup right the first time, we can make adjustments on the fly. Just get the machine back up." You've probably never heard these comments in your shop. Truth is, we rarely stop to calculate exactly how many resources we waste daily on poor or inexact setups—material, direct labor, handling costs, disposition costs, electricity, heat, and the costs associated with doing it over again (and again, and again), not to mention the real lost opportunity costs when the customers take their business somewhere else.

• *Excessive customer order lead times.* In many cases, we fail to consider setup time as an element of our quoted lead time. As a consequence, because history has shown that we don't really know what our actual lead time is, we elect to quote industry-standard lead times. Or worse, Marketing and Sales quote whatever lead time the customer wants just to ensure we get the job. That puts everyone behind the eight ball when Manufacturing fails to live up to the delivery promise, or spends all of the margin doing so. There is a better way—effectively managing setup times through proven setup reduction techniques like the ones we discuss in this chapter. Let's get started.

### Managing Lead Time

To deal effectively with the challenges of reducing lead time, we must understand its makeup. Total lead time typically includes:
• Order entry time;
• Order processing time;
• Order planning time;
• Material procurement time;

- Receiving, storage, and issue time;
- Material move time;
- Equipment setup time;
- Material run time;
- In-process storage and handling time;
- WIP warehouse storage time;
- Packaging and shipping time.

Does your currently quoted lead time include all of the above? If not, why not? Are they not real elements of the equation? Are they not time consumers that occur every day, for every customer order? Then why do we so often fail to accurately measure and include them in our calculations, versus pulling a number out of the air? Before we can begin to reduce our lead time, we must first understand and measure all of its elements. It just makes sense.

In the past, most of our emphasis has been on reducing the run time in our manufacturing processes. Manufacturing engineers for years have designed or purchased ever-faster production equipment—CNC centers, CIM systems, cells, etc. Those were valiant efforts, for sure, but were they timely? After all, should we not concentrate our efforts on the largest elements of the lead time equation first following Pareto's Law that 80% of the lead time is a result of 20% of its elements? If we were to do so, considering all of the elements of lead time, what percentage of the sum total of those elements do you think is run time?

- 75%,
- 50%,
- 25%,
- 15%,
- 10%, or
- 5% or less.

It perhaps may come as a surprise that in most industries, run time represents less than 5% of the total.

Another point: from the customer's perspective, other than run time, which of the elements of total lead time listed above can be considered value-added? Recent studies of American enterprises revealed that up to 95% of the activities related to a given process do not add value to the product or service provided by the organization, as viewed by their customers. In our example, that statistic indicates that only the actual transformation of raw materials into finished products (run time) actually adds value.

Traditionally, our efforts have focused on enhancing actual run times through a variety of technological advancements, while our Japanese competitors have concentrated on reducing or eliminating the nonvalue-added elements that comprise up to 95% of the total. The result has been a rapid growth in their ability to introduce and manufacture products faster than their American and Western European competitors, giving them the competitive edge in industries like electronics, capital equipment, HVAC, and automobiles. It's time to reverse that trend.

*Isolating the Nonvalue Elements in Lead Time*
If we look around the factory, many of the lead-time reduction opportunities become apparent. The manufacturing engineer must first isolate, then measure, and ultimately establish priorities relative to which of the nonvalue-added elements to resolve first. Listed in Table 6-1 are examples of commonly observed lead-time consumers that provide tangible opportunities for immediate lead-time reduction.

## WHY ADDRESS THE WASTE IN YOUR PROCESSES?

The benefits are obvious:
- *Survival.* Waste inhibits speed and flexibility to dynamic customer requirements and expectations. World-class competition forces us to react quickly to unexpected changes, or lose.
- *Profitability.* Waste costs money, big money. Remember the cardinal rule of business economics:

### Costs + Margin = Selling Price

If your costs are not under control, neither will your profitability. Why? You can no longer dictate selling price. Only the customer can. Thus you have only two choices: 1. Take it out of the margin or, 2. Cut your costs. Guess which one the boss will pick.
- *Quality.* Waste impacts the quality of your products and services, the quality of your work force, the quality of your processes, the quality of your customer service, even the quality of your environment.
- *Waste kills*! It kills the business, customer trust, employee morale, market share, profits—in short, everything you've worked for.

## Table 6-1. Clues to Lead-time Consumers

DEFECT RESOLUTION TIME resulting from:
- Nonconforming raw materials and components,
- Incapable and/or uncontrolled processes,
- Process spoilage or inefficiencies, or
- Rework, sorting, and use-as-is decisions.

QUEUE OR WAIT TIME that includes:
- Waiting for materials,
- Waiting for documentation or prints,
- Waiting for tools, jigs, or fixtures, and
- Waiting for the equipment to begin a changeover.

INSPECTION TIME in:
- Receiving,
- In-process, and
- Final quality acceptance.

EXCESS PRODUCTION TIME, that results in:
- Wasted material,
- Wasted labor, and/or
- Wasted capacity.

EQUIPMENT DOWNTIME, that results from:
- Unplanned or unscheduled shutdowns,
- Breakdowns,
- Excessively long setups, or
- Unchecked wear, vibration, and deterioration.

INVENTORY STORAGE TIME (often unnoticed as a time consumer):
- Stocking, handling, and moving of raw materials,
- Stocking of WIP awaiting further processing, and
- Stocking and handling of finished goods.

MATERIAL HANDLING AND TRANSPORTATION TIME, impacted by:
- Distance from storeroom to point of use for inventory and tooling,
- Distance from receiving to storage, and
- Any outside processing requiring the material to leave the plant once the processing has begun.

OTHER TIME consumers such as:
- Engineering changes,
- Bill-of-material or specification errors,
- Routing errors,
- Setup errors, or
- Secondary processing.

For example, by focusing on the reduction of waste in setups, the Japanese have rapidly moved into a position in which they can produce their products at a rate that matches their order intake. This allows them to provide extreme customization of their products while maintaining the benefits of cost and quality associated with traditional mass production.

Domestically, since the late 1980s, almost 30% of the metal stampers in the U.S. with annual revenues of more than $50 million have reduced setup times by at least 25%. In addition, more than 75% of those same metal stampers have reduced their total lead time by a similar or greater percentage. And, using quick-change tooling, the majority (again, more than 75%) of U.S. metal stampers have reduced inventory to comply with their customers' requests for JIT-type deliveries.

### Take Preventive and Predictive Maintenance

Today, world-class performance in the manufacturing sector dictates advanced planning of at least 85% of all maintenance activities to ensure a minimum of 95% planned up time on all essential production equipment.

A report in SME's *Manufacturing Engineering* sets the tone for the future for those whose intent it is to remain competitive in the manufacturing industry.[3] In the magazine, comparisons are drawn between current industry average performance and world-class performance in a number of categories. Condensed into tabular form, the comparisons are shown in Table 6-2.

**Table 6-2. Industry Average
versus World-Class Manufacturing**

| Metric | Industry Average | World-class |
|---|---|---|
| Setup times | 200 Minutes | 10-15 Minutes |
| Quick-change tooling | 26% | 85+% |
| Inventory turns | 9.2 times annually | 40+ times annually |
| Planned maintenance | 42% | 85+% |
| Manufacturing cycle times | 15 Days | Less than 2 days |
| First-pass yields | 95% | 99+% |
| Total downtime | 28% | 2 to 5% |

### How Do We Begin?

It is safe to say that virtually all companies are up against economic constraints. The best employment of resources, money, equipment, and personnel are, therefore, paramount to continued success. Why? Because there are only limited resources available to satisfy increasingly intense customer demands. Because of this, every manufacturer must maximize productive time and minimize those elements within all processes that introduce waste. Many times a change in culture and a change in thinking are required.

To effectively address the waste within our organizations, it's essential to first conquer the fear of and natural resistance to change. Change is hard to deal with. There's no doubt about it. But it is also a requirement for us to move on to the next plateau of competitiveness. Henry Ford once said, "Those who say it can't be done, and those who say it can be done are both right." There are many variations on the theme.

- It's not my job.
- You are trying to make my job harder.
- We've always done it that way.
- If it ain't broke, don't mess with it.
- Don't rock the boat.
- Management is just trying to eliminate more jobs.
- No one knows how to set up that job better than old Jim.
- I don't know how they do it, but it works. Leave it alone.
- You will never be able to improve the way we set up those processes; we've been doing it that way for years.
- Time is of the essence.
- We must insist on a 1000-piece minimum order to absorb our setup costs. The customer will just have to take what we offer.
- We can't do anything about our lead time. It is competitive with our competitors in the industry. What more could the customers ask?

These are the mindsets that must be changed before we can move forward into world-class competitiveness. For the remainder of this chapter, we concentrate on many of the practical methodologies the manufacturing engineer can use to overcome these mindsets and achieve quick results to the challenge of reducing setup and changeover times, work-in-process inventory investment, and the lead time to customers.

Our approach will focus on the 50% rule, with an overall objective of achieving the capability to set up for one unit. Is a single piece setup reasonable? Not in all cases—just as the concept of "zero inventory" is unreachable. But if we set our targets for the maximum reduction in setup time possible, and if we continue to focus on further reductions as technology permits, we will achieve an optimum balance between the ideal and the real.

### Freedom to Challenge

The change process requires the freedom to challenge the status quo. We have to accept new rules to govern the way we approach our jobs. We must:

- Understand why and how each element of our key processes occur.
- Challenge even established processes. Just because we have done them the same way for years does not mean they meet the needs of today's dynamic market requirements—nor does it mean that our prior methods were not proper for their time and their environment.
- Understand there are no sacred cows. Everything is fair game.
- Be conscious of the people involved—their fears and their concerns. We must solicit their active participation, because their active resistance will cause ongoing delays and problems, perhaps even failure.
- Assess the resources required and allocated to each process. For example, assigning two setup operators to a given setup will double the manning, but if the total setup time is reduced by more than 50%, the returns make the labor investment worthwhile.
- Evaluate the process metrics and control effectiveness. If the setup process is not under control, your efforts to improve it may not yield the desired results.
- Follow the principle of keeping it simple. Complexity builds cost and consumes excess time. And if you make the process difficult for the average setup operator to understand or follow, he or she will surely revert to the old methods—you can count on it.
- Assess the use of data capture technologies to enhance the speed of the processes. For example, bar code scanners are

now commonly used to scan work orders to identify which materials and fixtures are required for a given job and their current status. This is all vital data to have before beginning a setup process.

- Organize into a cross-functional team to address setup reduction opportunities. The axiom that two heads are better than one has merit. Employ the experience and talents of all players to your advantage—setup experts, operators, maintenance personnel, process customers, process suppliers, and even novices. They understand the process and its nuances better than anyone.
- Be open to suggestions and critique from the process owners.

*The 50% Rule*

The 50% Rule focuses attention on reducing setup and change-over times by 50% with each pass. In other words, if our current setup time is 4 hours, the 50% Rule would direct that we reduce it to 2 hours on the first pass. Then cut it to 1 hour—then to 30 minutes, 15 minutes, 7.5 minutes, etc. The focus is on continuous improvement.

Recognize that all setup reduction ideas are good and some are better than others. The best ideas are those that can be implemented at a low cost or at no cost. At first blush, that is typically hard to believe until you have some setup reduction experience under your belt. But it's common for a setup reduction team to achieve 50, 60, 75% reductions on the first passthrough—*without a major capital investment.*

The same principles used in core process re-engineering apply to the stages of a typical setup process.

- Identify how the setup process "actually" works.
- Define and isolate each step in the process.
- Determine the time and resources consumed by each step in the process.
- Assess the value of each process step.
- Isolate the waste, redundancies, disconnects, bottlenecks, etc.
- Re-engineer the process for optimum use of time and physical and human resources.
- Thoroughly document each step of the new process, its metrics, and its controls.

- Communicate the benefits of re-engineering to the process owners, then train them on the use of the new process.
- Start again.

Remember, *never automate a poor or inefficient process. You're just building in additional fixed costs.*

By definition, setup reduction is the process whereby the consumed time from the last good piece of the previous run to the first good piece of the following run (at normal operating conditions) on any production center is reduced without an adverse effect on quality or total process cycle time. In other words, effective setup reduction requires that a given process be optimized without an accompanying suboptimization of any other downstream process(es).

## HOW TO ADDRESS WASTE IN YOUR PROCESSES

### The Five Primary Elements of the Setup Process

The setup operation in nearly all manufacturing activities can be broken down into five primary elements.

1. *Getting ready*, which accounts for roughly 20% of the consumed time. This element consists of:
   - Obtaining the work authorization from the scheduler/dispatcher.
   - Delivering the materials from which the parts will be produced to the work center.
   - Accumulating the required documentation to produce, inspect, and test the parts produced (routers, drawings, Statistical Process Control [SPC] sheets, etc.).
   - Delivering the required tools, jigs, and fixtures to the work center and verifying they are in "ready-to-use" condition.
   - Delivering hand tools, fasteners, and any other required changeover tools to the work center to complete the changeover.
2. *Removal of the old tools and placement of the new tools*, which accounts for only about 5% of the start-to-finish (consumed) time.
3. *Locating and setting of the new tools*, which under normal operating conditions accounts for about 10% of the total consumed time.
4. *Preliminary tool adjustments*, which consume 15% of the total setup or changeover time.

5. *Final adjustments*, which account for the remaining 50% of the total consumed time.

From our earlier discussion of nonvalue-added time, which of the elements described earlier appear to add value to the setup or changeover process?

| Element | No Value | Value | Questionable Value |
|---|---|---|---|
| 1. Getting ready | ☐ | ☐ | ☐ |
| 2. Removal and placement | ☐ | ☐ | ☐ |
| 3. Locate and set | ☐ | ☐ | ☐ |
| 4. Preliminary adjustment | ☐ | ☐ | ☐ |
| 5. Final adjustment | ☐ | ☐ | ☐ |

### Identifying Work Activities

The secret to effective setup reduction is to first eliminate all elements of the process that add no value to the setup or changeover process. Thereafter, we have to concentrate on moving as many of the remaining activities as possible *external* to the setup or changeover process to minimize the time the machine is down. The object is to leave only those activities that must by accomplished while the machine is down (*internal*) to be re-engineered on the first pass. This is known as the Internal-External Method.

As described by Phil Stang and Jerry Claunch in their book *Setup Reduction . . . Saving Dollars With Common Sense*, most, if not all, setup activities can be defined as:

Internal: Work that *must* be done while the machine is shut down. and
External: Work that *can* be done while the machine is running.[4]

Obviously, the more external activities that can be performed, the shorter the changeover time because the external activities can be performed concurrent with the production of good parts. Our focus is, thus, to accurately classify all remaining activities after the value-added/nonvalue-added assessment as either internal or external, then to concentrate our re-engineering efforts on moving as many internal activities as possible to be completed external to the setup process.

The balance of this chapter is structured to guide you through the process of isolating and eliminating nonvalue-added activities, then re-engineering the remaining activities for the shortest total setup and changeover cycle times possible.

### Re-engineering the Setup

The first step in re-engineering the setup process is to define the "as-is" status, that is, how the setup is actually performed. This does not necessarily mean how the procedures say it should be done, nor how the supervisor says it should be done, but *how it actually is done today*. The approach most commonly used today to isolate every element of the process is with videotape, in which a video recording is made of the entire process from beginning to end—no cuts, no edits—just as it happens, including all activities and their associated timing.

The video recording approach, however, requires preparation, and the manufacturing engineer should consider the checklist in Table 6-3 prior to the taping session.

#### Table 6-3. Preparing for "As-is" Videotaping

| Confirmed | Not Confirmed | Preparation |
|:---:|:---:|:---|
| | | CAMERA PREPARATION |
| ☐ | ☐ | Are clock and date feature operational? |
| ☐ | ☐ | Are all controls and features operable? |
| ☐ | ☐ | Is a tripod available? |
| ☐ | ☐ | Are extra batteries available? |
| | | LOCATION PREPARATION |
| ☐ | ☐ | Is there room to operate the camera? |
| ☐ | ☐ | Is the total setup in view? |
| ☐ | ☐ | Is the lighting adequate? |
| | | TRAINING AND PERSONNEL |
| ☐ | ☐ | Is the camera operator trained? |
| ☐ | ☐ | Have all personnel been advised? |

*Filming the Setup*

As with the equipment preparation, there are several suggestions and guidelines for filming the setup process:

- Clearly explain to the operators and setup personnel what is being recorded and for what purpose. Get their participation and buy-in.

- If you are not familiar with the equipment, prior to the taping have the setup operator give you a general overview of the "as-is" setup including all the key activities. This will ensure that you include all essential movements and activities in the video.
- As a screen director would do before beginning to tape the process, frame the entire operation, machine, or operator in the camera. Later, you can focus on the specific parts of the operator's body that are involved with the setup or changeover activities.
- When you begin, carefully film all operator and helper movements. If they leave the general filming area, continue the video until they return to ensure all consumed time is accounted for. A brief narration of the reason for their absence from the viewing area is often helpful, such as: went to retrieve materials, looking for tools, went on break, etc.
- Review the video with the operators and setup personnel for confirmation that all setup or changeover activities have been captured on film. If you have missed any critical activities, it may be wise to capture them on a secondary video.
- Make sure the time and date of the video recording are visible on the film. These will be used during the assessment to establish the cycle time of each element in the changeover process.
- Review the film with the setup reduction team as part of the re-engineering process. During the review, ask the operator or setup mechanic to narrate the film if no narration was done during the actual filming to aid in the understanding of why and how each step was performed.
- The entire team should observe the film and participate in the re-engineering activities. All input is critical to ensure the best solution.

*Advantages of the Video Method*

The video methodology has a number of distinct advantages in isolating the "as-is" process flow. First, the video can be replayed as often as needed to ensure a complete understanding of all activities in the setup or changeover process. Secondly, the video approach allows the operator and/or setup mechanic to participate in the review to explain each step in the process, answer questions, and provide any added detail that may be required on key skill issues required by the process. Next, the study of the process in visual mode facilitates a better understanding because it eliminates the tedious task of documenting each activity on paper, then attempting to match it seamlessly to the next with no deletions. Another significant advantage is that people tend to be most objective when they observe themselves performing an activity versus being told what actions they have taken. As such, they are much less defensive and more open to the required re-engineering changes because they have been a part of the change process itself.

Open discussions during the viewing of the film offer still another advantage because they often lead to immediate improvement ideas that require little time and money to resolve. They become the "quick hits" that provide an immediate payback for the time and money spent by the manufacturing engineer and his or her team in preparing for the video process. The paybacks also provide working funds with which to implement additional setup reduction opportunities uncovered during later stages of the process. Remember, the video method is typically more readily accepted by operators and setup people than are the typical industrial engineering time studies because it provides direct versus second-hand data relative to individual actions and their results. Additionally, visual information is easier to understand and is retained longer than either verbal or written information, especially when studying complex and multiple-motion operations.

### Setup Re-engineering Worksheet
The worksheet depicted in Figure 6-1 is designed to provide you with the structure to:

- Compile the observed activities onto one form and assign times to each.
- Determine the total "as-is" cycle time for the process.
- Assess each activity for its value to the setup process.
- Separate internal and external activities.
- Re-engineer the process for improved cycle time.
- Estimate with a high degree of confidence the improved cycle time of the re-engineered process.
- Document the new process for further enhancement, implementation, and training of operators and setup personnel.

### Setup Re-engineering Worksheet

| Operation Number | Activity Description | As-is Time | Internal/ External | Value- added? | Eliminate or Re-engineer? Description | Target Time | Actual Time |
|---|---|---|---|---|---|---|---|
| | | | | | | | |
| | | | | | | | |
| | | | | | | | |
| | | | | | | | |
| | | | | | | | |
| | | | | | | | |
| | | | | | | | |
| | | | | | | | |
| | | | | | | | |
| | | | | | | | |
| | | | | | | | |
| | | | | | | | |
| | | | | | | | |
| | | | | | | | |
| | Total | | | | | | |

*Figure 6-1. Assessing and documenting all setup activities establishes baselines for determining whether or not to re-engineer the process.*

*Use of the Re-engineering Worksheet (Helpful Hints)*
Observations from the video should be tabulated on the worksheet in as much detail as possible. Operation or sequence numbers are used to differentiate one activity from another. A brief description of each activity should then be compiled (e.g., tightened lock nut, 15 turns; looked for wrench in tool box; returned to stockroom for additional setup pieces). The actual time for each activity should be recorded from the videotape, then totaled. This establishes the baseline for improvement. Next, the manufacturing engineer and his or her team should determine if the activity is internal or external, and if it adds value to the setup process.

The first pass reductions can now be determined. Those activities that are obviously nonvalue-added, like waiting for materials, looking for tools, waiting on prints, etc., can be quickly addressed. In a similar manner, those activities that can be easily shifted from internal to external are addressed at this juncture. The re-engineering comes next.

During the assessment of the existing activities, verify that each operation is neither overstated nor understated. Determine how much of the total cycle time is consumed by internal and external activities by noting each on the worksheet along with its corresponding cycle time. This is the separation stage of the re-engineering process. The objective of the separation stage is to isolate those activities conducted internal to the setup operations in order to focus our initial re-engineering activities in those areas. By so doing, we can minimize equipment downtime, one of our key metrics.

Typical opportunities to watch for during the separation stage include:

- Transportation of materials or products to or from the stockroom while the equipment is down.
- Movement of tools, jigs, or fixtures to the machine after the machine has been shut down.
- Discovering defects in tools, jigs, or fixtures after they are mounted in the machine and production has begun.
- Making equipment or tool repairs during the setup or changeover process.
- Transportation of tools, jigs, and fixtures to the tool room while the machine is down.

- Searching for tools and fasteners while the machine is down.
- Crawling under or behind equipment to access fasteners or tie-downs.
- Die heatup time once it is in the machine.
- Lost documentation for dies, jigs, and fixtures.
- Changeovers requiring more than one employee.

### Practical Solutions

There are a number of practical, low-cost solutions that can be applied during the re-engineering stage of the process to reduce the amount of equipment down time. There is no magic in any of these recommendations; they are simple back-to-basics solutions that can be applied to most operations in most manufacturing industries.

The first of these is the common *checklist*. Checklists are a good tool to use to avoid oversights and mistakes that typically arise once the setup or changeover has begun. The checklist includes:

- Listings of the required tools, materials, documentation, and procedures for a given setup.
- Listings of optimum speeds, feeds, temperatures, pressures, and similar settings for a given material or machining center.
- Listings of required resources and personnel for a given setup.
- Listings of the correct product and process metrics for each operation.

The second of the tools is a series of *functional checks* that can be used effectively to ensure all required items for the setup or changeover are functioning correctly prior to installation on the machining center. The objective is to correct any functional defects prior to the initiation of the setup or changeover process rather than after the machine is down. Functional checks typically include the required tools, jigs, fixtures, gages, and dies specified for the setup, along with any required hand tools, power tools, hoists, forklifts, or die tables.

An often overlooked back-to-basics step is the transportation of tools, jigs, fixtures, dies, and materials to their point of intended use before the changeover process begins. Movement of these items should be performed either by the operator while the machine is operating or by an indirect employee in cases where an operator cannot leave the machine.

The final step in the re-engineering process is to assess every remaining step in the setup or changeover process. The manufacturing engineer and his or her setup reduction team must adopt an approach that questions everything. Nothing should be taken for granted. Pursue questions like:

- What is the objective of this step?
- Why is this step being performed this way?
- Why is this step being performed at this stage of the process?
- Is there a better place or time in the process for this to be done?
- Can this step be eliminated altogether?
- Can any of these steps be performed concurrently?
- Why is this person(s) doing this?
- Why is it being done in this manner?
- How could it be done better?
- Does this really make sense?
- Can this step be simplified or combined into another step?
- Does what is being done add value to the customer?

Remember, you are looking for facts. Working with hard data is essential if improvements are to be made.

### Improving on Internal Activities through Re-engineering

The following additional hints and suggestions will help in your re-engineering efforts of the internal activities:

- Look for activities that concurrency can be applied to: for example, steps on which two or more people can work together to reduce the total consumed time by 50% or more. Remember, concurrent activities are 100% value-added. If the activities proposed are not, you are merely doubling the overhead for any given setup.
- Look for simplified clamping devices versus conventional bolts. Tools like pneumatic and hydraulic clamps provide quick, solid clamping, as do 1/4-turn fasteners that reduce securing time and effort.
- Use pear-shaped holes to minimize time for removal and replacement of fixtures and clamps secured by fasteners and bolts, along with U-shaped slots to eliminate the requirement for complete disengagement of tools and clamps. Also consider the

use of single-motion devices like slip-and-ball-lock pins to greatly reduce the time to secure and locate fixtures.
- Always use the right tools for the job, e.g., power drivers versus open-end or box-end wrenches, and permanent tools where space and safety permit.
- Above all, always strive to minimize or eliminate adjustments by using constant numerical values and scribe marks for visible centering.

### Fixture and Tooling Standardization

The use of standardized tooling and fastener sizes will further reduce setup and changeover time by minimizing the number of power and hand tools required to complete the process.

As an example, power press setups typically require adjustments for press shut height. The time lost resulting from constant raising and lowering the press plates to accommodate varying die sizes can be minimized using blocks to standardize the overall height from one die to the next.

### Centering Jigs

The alignment of a ram hole and shank typically requires tedious effort to "inch" the ram down to adjust the final die position, often by sight. Another, more time-effective approach is to attach a centering jig to the die attachment plate, thereby ensuring perfect alignment the first time.

### Re-engineering External Activities

Once the internal assessment and re-engineering activities have been completed, there are a number of questions the manufacturing engineer and setup reduction team should consider when beginning to re-engineer the external activities. It is not enough to just shift internal activities to external. We must make every effort to eliminate them altogether. If they cannot be eliminated, we must re-engineer the remaining external activities for better organization, control, accessibility, precision in measurement, lower cost, and reduced cycle time.

The team should explore the following questions:
- What controls must be in place and monitored?
- How can all required items be better organized for this setup?

- Where should tools, jigs, fixtures, materials, and documents be located to facilitate a speedy changeover?
  - What methods can be employed to ensure their availability at time of need?
- How can we ensure tools and fixtures are maintained in working condition?
  - How many of each are required to keep in stock?
  - Who is responsible for delivery and maintenance of these items?
- Are all activities that are being performed really required?
- Can any of the existing external activities be improved?
- Are the proper resources allocated to ensure a timely and efficient changeover?

### Setup Categories

Every setup can be dissected into four primary categories:

1. *Organization.* This category encompasses the gathering of all the necessary materials, tools, documentation, fixtures, jigs, and transport equipment required to perform the setup or changeover *prior* to the initiation of the setup activities versus during or after.

   The tools and techniques typically used to improve organization include checklists; functional checks on tools, fixtures, jigs, and materials to ensure 100% compliance with quality specifications; and the storage of tools, dies, fixtures, jigs, and materials in proximity to their point of use. For example, if a die is used 80% of the time on one press and 20% of the time on three others, it makes good sense to store the die near the first machine where it is used most.

   A simple, but effective technique I have used with great success is the placement of chalkboards or flip charts at each critical machining center. As illustrated in Figure 6-2, the chalkboard contains the status of the next three scheduled jobs for that center. As the required materials, tooling, and documentation are staged ahead of the center, they are noted on the board. As the prior run is nearing an end, it becomes a simple task for the operator to glance at the board to determine which of the scheduled jobs has all of the necessary materials, tools, and documentation available. The setup is

| Job | Materials | Tooling | Documentation |
|-----|-----------|---------|---------------|
| A | x | x | |
| B | x | x | x |
| C | x | | x |

*Figure 6-2. Shop-floor chalkboards keep operators constantly informed of the status of jobs scheduled for a particular work center.*

begun only on those jobs for which all of the required elements are available, thus eliminating the tendency to begin a setup, only to have it stalled waiting for one or more of those elements once the process has begun. Take a look around your shop. How often are setups halted while the operators head off to look for parts, tools, or paperwork?

Before moving on, answer some additional questions relative to organization. *Given that we are all engineers, and by nature very organized, would I be correct in assuming that a review of your personal tool box would indicate that all of your tools are neatly placed by order of size and configuration? Are all open- and closed-end wrenches placed sequentially in series, blade and Phillips screwdrivers separated, sockets separated by size and configuration, etc? Are they all clean and in perfect working order?*

The truth is, most technical people (including setup mechanics) tend to let little things like these slip. The result is that they spend from three to five times longer finding and retrieving their tools than would they have if the tools were correctly organized. While at home this may be no big deal— 5 minutes here, 10 minutes there—it can really add up during a setup operation, especially if several different tools are required. Next time you watch a setup, look to see how much time is lost hunting for the correct tools.

2. *Removal and installation.* This second category involves the removal of the old tools, dies, jigs, or fixtures and the placement of the new tools, dies, jigs, or fixtures on the machine. This typically takes the shortest amount of time of the four categories.

Several of the improvement techniques for this category were mentioned previously. Things like:

- Using 1/4-turn fasteners, pneumatic and hydraulic clamps, and single-motion fastening and clamping devices,
- Reviewing and removing all unnecessary torque requirements in an effort to eliminate a secondary operation and additional tooling and calibration requirements, and
- Using die carts, tool carts, die transfer lines, intermediary jig and die plates, and rotary tables.

3. *Setting the tools.* This category involves locating the new tools in proximity to their final points of use on the machine. This, too, normally represents a small percentage of the total setup time consumed.

Tools and techniques used to facilitate locating the new tools typically include centering jigs; visible (scribed) centering lines for quick, visual locating; constant numerical dial-in values that have been predefined from prior optimization efforts; and locating pins and centering chucks (Figure 6-3).

4. *Adjustments.* Adjustments should be minimal, quick, and simple as we discussed previously in some detail, though they

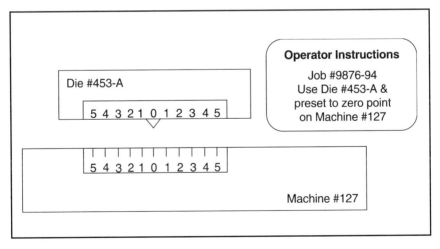

*Figure 6-3. Simple, on-machine aids can cut setup time by helping the setup person locate new tools.*

normally represent as much as 65% of the total consumed
setup and changeover time. This category includes:
- The initial adjustments to start the process (just getting
  it close);
- A check of the results of the initial settings;
  - Dimensional,
  - Temperature,
  - Pressure,
  - Volume attainment, etc.
- A secondary adjustment to bring the initial adjustment
  to the specified requirement; then
- A final check of the results of the secondary adjustment
  to be sure it meets spec, and so on. This cycle continues
  until the specification has been achieved and production
  has attained the desired rate of efficiency.

When these adjustments cannot be eliminated altogether,
a number of tools and techniques can be used to enhance this
category, including the use of preset stops and positioning
devices; application of standardized settings taken from optimum run conditions that have been documented and verified; locating pins and slots for zero-tolerance fits; and the
assignment of skilled, trained operators and setup mechanics to complete the setup.

### And Then There is the Documentation

What's available? What's needed? In most companies, the setup
processes have never been documented. The combination of the
videotape and corresponding setup re-engineering worksheet, however, provides an excellent baseline for the development of the
needed documentation. Remember, the documentation should always be dynamic. As process improvements are implemented, they
should be incorporated into the setup documentation for use by all
parties.

The ideal setup documentation should include:
- *Current specifications* regarding the products to be produced,
  along with corresponding process routings, inspection and test
  requirements, and any special process instructions.
- *Quality assurance and process control data* to denote the frequency of audit/inspection, confirmation methodologies, SPC
  frequencies and limits, and gages and checking fixtures to be

used, along with the frequency and methodology of their calibration.

- *Equipment maintenance procedures* and frequencies, predictive maintenance results, and corresponding corrective actions for all dies, jigs, fixtures, and tools.
- *Tool crib procedures* that define issuance requirements and stocking levels for critical setup materials and tools.

As mentioned earlier, the maintenance of all setup and related documentation is essential. So why is it so frequently overlooked? Simply put, no one is assigned the responsibility of maintaining the records. No one is assigned the responsibility of overseeing the maintenance of the records to ensure it is done. No one is assigned the responsibility to audit the setup procedures and records. No one is held accountable and, predictably, nothing gets done. And let's face it, paperwork is considered by most shop people as something to put off as long as possible. It just isn't considered important. That is a mindset the manufacturing engineer must overcome if significant setup improvements are to be realized and continued.

The most effective method of ensuring that the setup records are maintained is to tie the documentation responsibility to the person performing the task, either the operator or setup person. Make the ownership theirs.

## INTEGRATING QUICK CHANGE INTO THE DESIGN PROCESS

Today, broad-based enterprise objectives are being used to focus all corporate disciplines on reducing the time it takes to get a product to the customer—the total time, not just the actual production time. To aid in that reduction process, both product and process designers have been enlisted to design products and processes that can be rapidly converted from one to another. Using concurrent engineering techniques, designers, manufacturing engineers, industrial engineers, tool designers, production operators, and setup personnel collaborate on the design of the product, the processes that will produce those products, the tooling that will be used, and the changeover requirements to meet dynamic customer expectations—all before the design is reduced to paper or finalized through

CAD. The benefits of designing in rapid changeover capabilities are obvious:
- Creation of only value-added labor and overhead costs,
- Minimization of tooling procurement and maintenance costs,
- Lower material and scrap costs,
- Lower rework costs,
- Shorter cycle times,
- Increased agility and flexibility to customer changes,
- Reduced lead times to customers,
- Lower inventory investment, and
- Higher margins.

***Design Guidelines That Can Aid in the Changeover Process***
Several of the same quality design rules discussed earlier to facilitate producibility and assembly can be applied equally as well as aids to the setup process. Of particular importance are the following:
- Minimize adjustments (if they can't be eliminated),
- Design for single-surface processing,
- Reduce component and subassembly counts,
- Eliminate hand tool requirements,
- Eliminate fasteners or minimize their variations,
- Employ modular component concepts,
- Design for group technology,
- Design for in-process testing versus final inspection,
- Design for single-plane insertion,
- Design for top-down assembly,
- Design out hidden pockets and hard-to-reach features,
- Design for snap fits or "puzzle" fits, and
- Design for multifunctionality in parts.

## THE ROLE OF PREVENTIVE AND PREDICTIVE MAINTENANCE IN REDUCING SETUP CYCLE TIME

In most cases today, maintenance is an ignored function, addressed only when something breaks down. Most companies claim to have a preventive maintenance (PM) program in place. What they fail to consider, however, is that the existence of a PM program means nothing if it is not religiously followed every day, week in and week out. If the dies, jigs, tools, etc. are not in perfect working order, the setup will take longer to complete. It is therefore mandatory that critical production and changeover tooling be maintained in excellent condition at all times.

Several of the secrets to ensuring effective maintenance planning and execution are:

- Prior to the completion of production, a final-piece inspection should be performed to ensure that the dies, tooling, jigs, and fixtures are producing product that is in 100% compliance with quality specifications.

- Based on the condition of the tooling *and the results of the latest predictive maintenance activities*, the tooling should be returned to its designated storage location, marked with a green acceptance tag carrying the date of inspection and number of impressions or strokes (if applicable), or red-tagged and returned directly to Maintenance for repair prior to the next scheduled use of the tooling.

- If predictive maintenance indicates that the useful or predicted life of the tooling falls within the normal run quantities anticipated for the next run of the product, the tools should be red-tagged and sent to Maintenance, *even if the inspection shows that the tooling is producing acceptable parts*. Why? Because the PM data shows that sometime during the next run, the tools will have deteriorated to the point that they will begin making bad parts. Remember, prevention is much less costly and time consuming.

## THE CLOCK IS TICKING

The need to address time reduction in setup and changeover is driven by customer requirements and internal requirements such as capacity limitations, time-to-market pressures, and customer requirements. The customer's focus is invariably on *time* and *cost*. If we decide to wait for the customer to force us into action, or for our market or industry to direct our activities, we will always be in a reactionary mode. Every day will be another fire drill. Nothing will be accomplished, and our focus will remain short-term.

The strategic decision to employ setup reduction must be generated from an understanding of why it is essential to sustain our core businesses competencies. The strategic decision to deploy setup reduction techniques must come from an enterprise-wide desire to enhance customer service and remain competitive in our markets.

### Opportunity Areas

Where do we begin? We begin with hard data. The accumulation of valid, verified data is essential to ensure that we have accurately focused our limited resources on those opportunities that will provide optimum return on investment.

There are many opportunities in the shop and the support areas to start with. Discussed in the following paragraphs are some of the principal areas in which to look to begin your assessment.

### Centers and Process Lines with Excessive Setup Times

This is normally the first area most setup reduction teams address because it is the most obvious. However, recognize that centers or process lines with the longest setups may not always represent those products that are in highest demand. As such, the investment in resources may not provide the payback you are looking for, or that is available from other reduction opportunities. As an example, think about those old service parts that are ordered only once a year. The setup is long, owing to the aging of the tools and the infrequency of their use, and the ROI is certainly not as great as it likely is on others.

Before selecting a setup project, carefully assess the data. Look particularly for the order frequency. Combine the order frequency (not lot size) with setup time to determine if the opportunity under investigation is significant. Use a Pareto analysis to segregate those few opportunities (typically around 20% of the total) that generate the majority of the setup hours consumed.

### Critical and Backlogged Work Centers

Even when capacity is in line with customer demand, at times certain work centers will become overloaded. These overloaded work centers can thus become a focus of the setup reduction team's investigation.

By reducing the setup time, the nonproductive downtime of the center can be reduced, bringing the center's practical capacity back up to the required demand levels.

### Significant Product and Process Lines

A product or process line that accounts for a significant percentage of the operation's revenues (50%+) may be a prime candidate for setup time reduction attention.

As before, a Pareto analysis will aid the team in selecting those products or lines that should receive the initial attention. Then, all components and subassemblies within the selected lines should be reviewed as well, again to determine through analysis which should receive the initial reduction thrust. Attention in this area provides a two-fold benefit: first, it reduces the process lead time so your customers can get their products sooner, and secondly, it increases your cash flow and revenue stream because your production and overhead costs are reduced and you get a faster return on your invested capital. Simply put, your customers get their products faster so you, in turn, get your money faster. Now that's a winning combination!

### Surplus Inventories
Other excellent targets for the setup reduction team are areas in which large surpluses of inventory are generated as a result of the need to absorb long setup times.

As setup and changeover times are reduced, order quantities can be reduced as well. This, in turn, will reduce the average inventory levels and increase the operation's inventory turns, a measure of utilization of invested capital. Remember, the cost of carrying inventory averages from 24-36% per year (over 2-3% per month). In addition, as lead times are reduced, the forecast accuracy will improve proportionately, thus minimizing your risk of producing and stocking products that do not conform to either the customers' requirements or anticipated needs.

### Frequent Stockouts
Another symptom of capacity and excessive setup time problems is frequent stockouts of critical components produced internally. Data will show whether or not the shortages are precipitated by raw material availability, setup, or quality factors. We frequently lay the source of the problem on unanticipated short lead-time customer demands. But business today requires us to be in a position to respond to those market and customer opportunities quickly and cost effectively. Thus, we can no longer blame the customer for our own inability to respond. Nor can we afford to hide our deficiencies under a mountain of safety stock.

Applying a Pareto analysis to the frequency of shortages of critical components will provide the direction for the team's initial investigations. Remember, the solution may involve a series of setup or changeover activities to ensure that total component or process cycle time is reduced to its minimum.

### Tool and Fixture Deterioration
Tools, jigs, fixtures, dies, process equipment, etc. whose lives have gone beyond a normal range (especially if preventive and predictive maintenance have not been performed regularly) are typically prone to longer and more frequent setups. The quality of incoming raw materials often plays a major role in early tool deterioration and should be a consideration in the team's corrective action plans.

### Don't Re-invent the Wheel
The setup reduction team should make an effort to document its activities and corrective action processes, and the results it obtained. For one thing, documentation supports the team's return-on-investment calculations. For another, it allows the team to replicate many of the corrective actions in other areas of the shop, thus reducing the implementation timing and associated costs.

Setup reduction teams should be encouraged to meet regularly with each other to share their solutions, experiences, and even failures. The faster the setup reduction methodologies and techniques are spread throughout the organization, the faster will be the cultural transformation required to sustain continuous improvement activities.

The best people for the team are those who have direct involvement in and responsibility for the setup and fabrication of the products. These include the operators, maintenance personnel, tool and die technicians, and setup personnel. These people are best because they know and understand the problems involved on a first-hand basis. On the other hand, simply because they *do* have that level of knowledge, there may be a tendency for those individuals to feel there is no better way to perform the setup than the way they have been doing it for years. Training will aid in overcoming those habits, but may not be enough to ensure an optimum solution. To get a broader perspective relative to the problem and its causes—and the ultimate solution—other personnel should be included on the setup reduction team.

## RESISTANCE

There will be those who will not be as enamored with changing their habits as you are. Resistance to change can almost be guaranteed. Where will it come from? Why will it occur? It will come from the supervisors, engineers, union leadership (if applicable), setup personnel, and operators—especially if they are not involved in the reduction process from day one.

The reason for resistance typically is fear and a lack of understanding of the process itself and of the principles and purpose of setup reduction. Resistance also stems from the transfer of authority for the setup activities from supervision to operators and setup personnel, the people closest to the operation. There is always a natural resistance to change, especially if the process has been in place for a number of years. Fear of the unknown forces people to naturally withdraw into their own comfort zones. Finally, if the union leadership is not involved as part of the reduction efforts, the traditional skepticism and distrust of management's motives will ensue. Is it management's intent to reduce the setup and changeover times so that they can reduce the hourly work force? Make us work harder? Cut back on our overtime? etc., etc, etc. Plato summed up the human dimension of change quite succinctly: "Man changes for one of two reasons . . . hope or fear."

Other issues include how the setup reduction team is structured, how training requirements are identified and the necessary instruction obtained, and determination of how long the reduction efforts will take and how much they will cost. All play a role in the cultural change process and, thus, must be considered.

## IMPLEMENTATION PLANNING

Planning the implementation of your setup reduction time process must be thorough and incorporate the depth of analysis to address all predictable and anticipated issues. At a minimum, the planning process should include:
- Development of the project justification criteria,
- Organization of the setup reduction activities and the players,
- Communication of the process to those involved and impacted by the reduction process,
- Training of the setup reduction team,

- Establishment of performance metrics,
- Methods of monitoring and communicating results.

The out-of-pocket costs associated with the process of setup reduction are initially minimal: training, video supplies, minor jig and fixture modifications, changes in fastening devices, storage and conveyance vehicles, etc. But remember, all investments must have an acceptable return in order to be justifiable to management. That is why we emphasized earlier the importance of capturing the baseline operating times. From them, we can establish a baseline cost and the point from which improvements such as the following will be measured.

- Increased output,
- Less downtime,
- Higher WIP inventory turns,
- Lower total inventory,
- Less unplanned downtime due to maintenance,
- Increased sales volume,
- Shorter quoted lead times to customers, and
- Lower setup waste costs.

Remember, *always look for the "low-cost, no-cost" solutions* to each setup problem. Also, bear in mind that all performance metrics must be quantifiable in dollars and/or hours to be truly effective. Management is focused on the bottom line. Any metric that is not visible there, isn't a good one.

It is imperative that the entire organization be aware of what steps were taken and the effort required to achieve results. Periodically (quarterly at a minimum), the progress made, achievements realized, and methodologies employed by the setup reduction teams should be presented to the operating and support organizations. Monthly presentations to the Executive Committee will keep the leadership abreast of the teams' accomplishments and will act as a vehicle to ensure continued support as ongoing improvement ideas are presented.

## DIFFUSING RESISTANCE TO ENSURE SUCCESS

As mentioned, resistance comes from a lack of understanding. One of the best approaches to overcoming resistance is to provide the training to enable those involved to understand the basic principles of setup reduction. By broadening their perspectives on new approaches and methodologies that can be employed to improve their individual jobs, your chances for broader acceptance will be greatly improved.

A risk assessment questionnaire is another tool that can provide excellent insight into the degree of resistance within an organization and the areas in which that resistance is greatest. The level of resistance can then be used to assess the degree of risk involved in implementing the proposed setup reduction activities, thus leading to identification of the best approaches to reduce the risk of failure.

## SUMMARY

There are no shortcuts to setup time reduction, no easy ways to achieve improvement without a directed approach that focuses the manufacturing engineer's attention on the solution to those problems that prohibit us from reducing our total cycle time and the corresponding lead times to our customers.

Setup is a process, and every process can be broken down into steps that can be streamlined, refined, and simplified. The important element of the equation is the human factor, getting buy-in at all levels—from board room to shop floor to union local—to head off to the extent possible, resistance to change that will inevitably emerge.

### References

3. *Manufacturing Engineering*, "Industry Notes," p. 10; April 1993. Society of Manufacturing Engineers, Dearborn, MI.
4. J. Claunch, P. Stang; *Setup Reduction . . . Saving Dollars with Common Sense.* PT Publications, Inc., Palm Beach Gardens, FL, 1989.

# 7

# Project Management

## THE PROJECT MANAGER'S ROLE

Project management is no longer merely a fill-in activity for manufacturing engineers. It requires a high degree of commitment to operational and fiscal results, an acceptance of accountability for conformance to project requirements, and the people skills to forge a synergistic chemistry between diverse functional groups.

Project management is a structured methodology for the analysis and management of complex assignments or problem-solving activities that are impacted by a multitude of variables, influences, and environmental factors. To be successful, the manufacturing engineer must break down each complex situation and condition into a subset of manageable, well defined activities that can be controlled and monitored to achieve established project objectives and expected deliverables.

What we will explore in this chapter is an approach that focuses on:

- Preparation for and acceptance of project management responsibilities;
- Isolation of project requirements and expectations;
- Selection and organization of the project management team;
- Establishment of the project scope, guidelines, deliverables, and metrics;
- Selection and utilization of the fundamental project management tools for decision-making, problem-solving, and situational assessments;
- Gathering data with which to make decisions, and the conversion of that data into usable information;
- Establishment of a project schedule and budget;

- Identification and assessment of alternatives for corrective actions;
- Tracking of costs and returns;
- Assessment of risks; and
- Preparation and sale of the proposed actions and solutions to senior management.

## ACCEPTANCE OF RESPONSIBILITY

Project management isn't simply giving orders and expecting employees or peers to blindly follow your lead. As a project manager, the manufacturing engineer typically has no direct line management position over the project team members nor the organizations impacted by the project, and thus very little real power. The manufacturing engineer has assumed or been given a great deal of responsibility and accountability, but is rarely granted the position with which to force change through the organization. As a result, you must employ a different approach.

Take, for example, the project team. To whom do the individuals report? To whom are they loyal? Who controls their paychecks? Get the picture? You cannot *manage* people who do not report to you; it simply doesn't work. Rather, you must *lead* them. That means you must be both accepted and trusted by them. And that dictates a different set of skills than are typically identified with the manufacturing engineer.

Today, project management requires extensive technological and facilitation skills, along with the capability to comprehend and develop strategic organizational objectives and the tactical skills to ensure their successful implementation. In addition, project management requires the skills to accurately assess customer (internal and external) requirements and expectations, and then to effectively prioritize them in order to establish project metrics. Organizational skills are required to select, develop, and motivate personnel from diverse functional and organizational groups to ensure that they will work together to attain established project objectives, and develop and implement project control processes to guarantee that the project remains on schedule and within budget. Project management also requires the follow-through to ensure that corrective actions are taken quickly and decisively, thereby guaranteeing that project objectives are never compromised. And, finally, project management requires the skills to know when and how to make decisions that will impact the overall project performance, costs, and schedule.

## THE PEOPLE PERSPECTIVE

To be an effective project manager, you must understand people. You must know how to isolate the multitude of motivational triggers inherent in every team member and have the skills to use them effectively. You must understand the importance of establishing attainable milestones and the associated metrics to monitor progress relative to them. You must establish rules and guidelines for the project management team to operate within, then hold every team member equally accountable for working within those parameters. You must earn the trust of every team member. To do so, you must be objective, consistent, and fair. Your project team must be able to count on you to follow the guidelines you have set, to stand up against confrontation, and to shield them from political influences that could hinder the successful completion of the project. The team members must be convinced that you will make the right decisions, and stand firmly behind those decisions.

Equally as important, you must understand yourself. You must recognize and accept your own weaknesses, as well as your personal strengths. You must endeavor to create synergy by building a project management team that complements those strengths and offsets your weaknesses. Accept the fact that your position as the project manager will require you to make tough, sometimes unpopular decisions.

Style is important. But because management styles are unique to individuals and the particular environment in which they are operating, they are not always easily replicated. Adopt a management style that fits you personally, as well as the scope and constraints of the project you have been assigned to manage. Remember, there is no one-size-fits-all when it comes to project management. The key is to lead by example.

Prepare yourself *emotionally* for the stress that inevitably comes with a project management assignment. Prepare yourself *physically* for the long hours and grueling pace that will be required. Prepare yourself *politically* for the turf battles that lie ahead. And prepare yourself *socially* for the selling that will be required to push your project up through the hierarchy of senior management approval processes.

### Some Guidelines to Follow

Before you begin, clearly understand what is expected of you and the level of empowerment you have been granted. In other words, what authority do you have, and what issues or decisions must be discussed with higher levels of management? What guidelines and rules have been placed on you in this assignment? How will success be measured? What project requirements are *musts*, and which are *like-to-haves*? Have all expectations been clearly defined—both implicit and explicit? Are they achievable? What are the risks (to the organization, to the project, and to you personally)? What are the rewards? From where is resistance likely to come, and to what degree? Who will mentor the project to aid in overcoming political resistance from the top?

Carefully gage senior management's commitment to your project. Measure that commitment in terms of:

- The level of resource allocation to successfully complete the project within the allotted time frame,
- The deployment of those resources when and where they are needed,
- The amount of capital budgeted for the project and the ability to allocate those funds without redundant approvals,
- The level of priority given to the project, relative to other projects currently under way or planned—*and* relative to the day-to-day operations within the organization, and
- The relative impact of the project on the critical or strategic business issues facing your organization.

Remember, it's great to be a hero, but only as long as you are around to receive the glory. Pick your battles carefully. Take project management assignments only if you have the ability, capability, and energy to be successful.

## SETTING THE PROJECT BASELINE EARLY

Why do we typically do such a poor job with project management? The reason is simple: we too often fail to isolate the project and customer requirements before starting; we make unfounded assumptions versus obtaining facts; we fail to quantify and prioritize the customer or project expectations; or we put blind faith in the expertise and direction of those above us. Consequently, we are constantly changing directions once into the project, changes which cost us time and money.

Before you commit, sit. Sit with the project sponsor and/or project customers and follow the guidelines in the model illustrated in Figure 7-1.

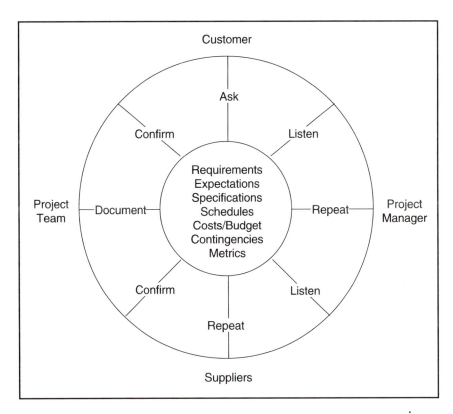

*Figure 7-1. By involving all affected parties in defining project parameters and establishing expectations, chance of misunderstanding is substantially reduced.*

*Ask* for definition and details relative to the project requirements and expectations. Then *listen* closely, taking notes of exactly what is said and how it is said. *Repeat* the definition and details exactly as you understand them, and ask for clarification. Then, *listen* again closely, making any necessary adjustments and modifications to your notes. Again, *repeat* the definition and details exactly as you understand them. Request *confirmation* of your understanding. If complete agreement has been reached, then *document* the project requirements and expectations for future reference by all parties. Then, again request a final *confirmation* from the project sponsor or customer.

By following this simple model, project requirements and expectations can be accurately defined in the beginning, thereby minimizing the number of changes and their impacts on the project cycle time, budget, and performance.

In defining the project, a comprehensive description is necessary to establish the scope, direction, and focus for the team. In addition, the intent or purpose of the project should be clearly delineated for the team to define the level of urgency or criticality of the project to the organization. The benefits to be derived from the project should also be described and quantified, along with any competitive issues. And, as with all elements of business, the metrics that will be used to gage progress and project performance must be clearly defined as part of the overall project description. It is wise to carefully consider all project expectations before committing to a baseline. Avoid being overly optimistic or overly pessimistic. Focus on achieving a realistic target, with continuous improvement thereafter.

Also essential to the success of the project at this stage is a comprehensive description of the project guidelines:
- What *is* and *is not* expected?
  - By when?
- What is the scope of the project?
- Are any of the project objectives or expectations in conflict?
- What are the rules under which the project team must operate?
- What operational constraints exist?
- What qualifiers must be considered?
- What are the budgetary constraints or considerations?

Defining what is not to be done is as important as defining its counterpart. The assessment of those issues that are *not* to be included within the scope of the project prevents inclusion of unnecessary project activities that introduce additional costs and resource consumption to the project. As such, duplication of effort and redundant activities are avoided. Project limits become clear and assumptions regarding the project are validated.

An effective approach is to involve the project sponsor or customer in defining the project:
- Scope,
- Deliverables,
- Schedule,
- Budget and costs,
- Performance expectations, and
- Metrics.

Involving the customer or project sponsor in establishing these elements of the project will ensure their ownership of the project prior to project launch.

## ESTABLISHING PROJECT METRICS

As stated throughout this book, metrics are critical to the accomplishment of project and business objectives alike. Simply put, that which is not measured will not change or improve. In setting metrics for your project, remember that the project will be judged by:
- *Performance*: the quantifiable, measurable results achieved from the project, and
- *Perception*: the subjective view of the effectiveness of the project in meeting the requirements and expectations of the customer or project sponsor.

Effective project management addresses both by answering the following questions:
- What will be measured?
- How will it be measured (methods, tools, procedures, etc.)?
- How often will it be measured?
- By whom will it be measured?
- What data will be required?
- What data collection systems will be required?
- What data conversion systems will be employed?
- What data system compatibility issues must be addressed?
- What reporting mechanisms will be employed?
- What will be the reporting frequency?

An assessment of the data collection systems is also in order at this stage of the project. The assessment should include an evaluation of the effectiveness and accessibility of the current data collection systems, along with a comparison of what they measure against what the project requires. Also of concern should be the timeliness of the data collection systems: are they real-time or batch processes? Without timely, accurate data, decisions cannot be made and progress against objectives cannot be measured.

When the project has been clearly defined, including its metrics and data collection systems, it is time to reduce everything to writing. This final step is critical as it formalizes the project baseline for all concerned. The project documentation should include the project definition and its deliverables, all performance specifications, project schedules and budgets, project costs, resource requirements and the timing of their use or consumption, project metrics and associated data collection systems and methods, reporting mechanisms and their frequencies, decision and corrective action procedures, the assigned level of empowerment, and all approval requirements.

*Now* it's time for everyone to sign on the bottom line—project manager, project sponsors or customers, any critical suppliers to the project, and the appropriate members of senior management. Project ownership is *now* mutual, so *now* you can begin.

## ORGANIZING FOR SUCCESS

Today, customers expect more from suppliers of both products and services: quality approaching zero defects; new products and services at an increasingly faster pace; a higher degree of customization to their unique environments or applications; smaller quantities at no premium in price; and products that are environmentally benign so that their use and disposal will not negatively impact the environment. Today's technologies provide the ability to do more for the customer, but no manufacturing engineer can keep up with the increased complexities of market dynamics and technological innovations alone. It simply isn't possible.

### Project Teams
To be effective, the manufacturing engineer in a project management role must employ all the resources available to him or her to their maximum potential. That means forging the human resources within the organization into an effective project management team to achieve the benefits of cross-functionality and synergy. The use of project teams has multiple benefits: first, they typically generate

more, and often better, ideas; secondly, cross-functional project teams are generally more creative in developing cost-effective solutions to operational and performance problems; thirdly, because of their numbers, project teams are better at analyzing large amounts of data and complex situations in shorter time frames. In addition, the perspective of cross-functional project teams tends to be more organizational than functional, focusing on what is best for the customer versus what may be best for individual departments or employees. Decisions focus on optimization of all applicable business functions versus the optimization of one function at the expense of another. And communication is enhanced because everyone is represented in the final solution or recommendation.

In many situations throughout industry today, however, project teams have proven ineffective. Why? Simply because most employees placed in a team or decision-making situation lack the skills and experience to function effectively. And because most organizations fail to recognize this, many teams are organized and sent into action without the least understanding of how they will achieve their goals, much less work together as a unit. Remember, most teams are organized around individuals with little, if any, formal managerial training or experience with which to make project-level decisions. The team members have more often than not spent their careers following directions from above and fearing the consequences of failing to follow those directions to the letter. So creativity, synergy, and the ability to "think outside the box" are typically not part of the team's readily available skill sets.

One of the keys to successful project management is the selection, direction, and motivation of the right people for your project management team. It's important, then, to spend some time examining several of the techniques used by professional project managers and business executives in building an effective project management team. Guidelines are given in Table 7-1.

Remember, to be successful, your project team must be a cohesive unit that works together. To create that environment, you must build a sense of "trust" first—among the team members and in you as their leader.

Trust, as illustrated in Figure 7-2, is derived when the project manager develops a vision for his or her project, then selectively recruits team members, with chemistry being a primary factor. As the chemistry among the project team members begins to take hold,

**Table 7-1. Guidelines to Effective Team Selection and Direction**

- Select individuals who have the ability to contribute to the project team as a "team member," not as a superstar;
- Select individuals who can contribute complementary skills that will increase the strength and capabilities of the project team;
- Select individuals who can perform under pressure and who are results-oriented;
- Select individuals who can both lead *and* follow;
- Select individuals who are committed to testing the boundaries of the status quo, who thrive on thinking in radical, unconventional terms;
- Select individuals who are willing to learn new skills from other team members and at the same time share their own expertise with those individuals;
- Avoid selecting politicians; let them run for office elsewhere;
- Select individuals with both technical and "people" skills;
- Select individuals from across the organization and top to bottom;
- Develop depth on the project team through cross training, managerial and technical skills training, and interpersonal creativity through intergroup mentoring;
- Establish rules and guidelines for decision making, problem solving, group interactions, dispute resolution, and assignment completion that apply equally to all team members; then hold all team members accountable for policing themselves and each other relative to these rules;
- With the project team, establish realistic project objectives, metrics, schedules, budgets, resource needs, expectations, and project deliverables;
- Control external and internal politics—shield the team from negative influences that could disrupt team harmony and trust;
- As the project manager, always communicate clearly and concisely;
- Being the project manager does not confer the right to micromanage; back off and let the team manage itself under steady-state conditions, and focus on facilitation versus management whenever possible;
- Be consistent in direction, approach, and scope;
- Set priorities, then maintain them unless conditions make a change essential to the success of the team's objectives or mission.

trust grows. Thereafter, as trust becomes implicit, the true synergy of the team becomes apparent in performance results.

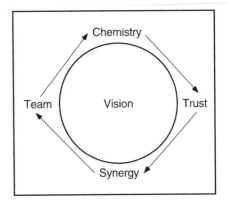

The establishment of that initial project vision is essential, as it sets the overall scope and direction of all future team activities. A clear vision is much like a puzzle (Figure 7-3). When all the pieces are in place, the picture is clear and easily understandable. But if pieces are missing, then the picture is incomplete; even distorted.

*Figure 7-2. Vision coalesces the components of team activity.*

The "pieces" of the picture, as illustrated in Figure 7-4, are the members of your project team. Each has a unique set of skills, background, and experience. Each brings something different to the project. Each is critical to forming a synergistic whole. If you select team members that don't "fit," your vision for the project will never materialize. And your project, and its intended deliverables, will never meet the intended expectations.

One last comment. A team is a reflection of its creator. As such, your team mirrors your abilities as a leader, as an organizer, and as a team member. As the project manager, the ball is squarely in your court. Handle it well.

A good place to start is by isolating the needed skill sets for team members to ensure compatibility with project requirements and expectations. Figure 7-5 is a sample matrix used by professional project managers to structure their search for the right combination of skills and experience.

Once the essential skills have been identified, the next step is to match the required skills against the corresponding project activity in which those skills will be required. In addition, it is now appropriate to begin identifying potential candidates for those critical skilled positions on the team.

The selection of candidates is often filled with bias and subjectivity if left to one individual. It is therefore best to seek the advice of several people outside the immediate group such as the project sponsor or mentor, human relations professionals, peers, and ex-

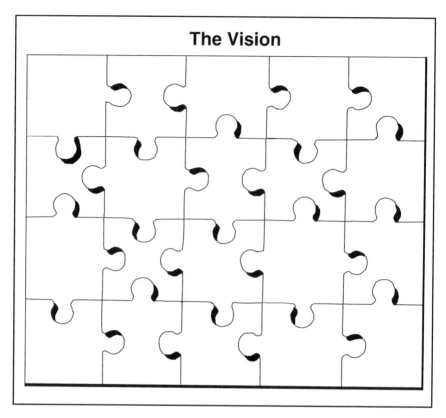

*Figure 7-3. Unless all parts of a whole dovetail, the overall picture, or vision, will be confusing and open to misinterpretation.*

ecutive staff members. Be open to differing opinions concerning candidates. It is helpful to create a skill needs matrix such as that shown in Figure 7-6. Always seek the most qualified candidates, considering both technical and people skills. And, *don't forget the most important ingredient in building a successful team: chemistry.* All the skills in the world will be useless if the members of the project management team lack the ability to work together.

Next, isolate the training needs for the individual team members to ensure their skills are up-to-date and their thinking is contemporary. When planning for team training, utilize training professionals who can bring real-world experience to the sessions. Again, a matrix check-off is helpful; see Figure 7-7. Theory is good only when coupled with hands-on applications training which

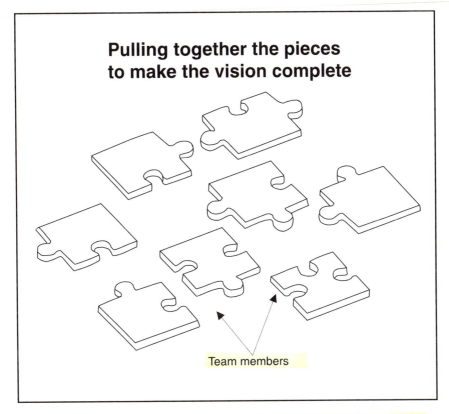

**Pulling together the pieces
to make the vision complete**

Team members

*Figure 7-4. The parts, or members, of a team, functioning as a unit, are more productive than the same members working individually.*

teaches the team members how to apply the techniques within their own functions and project activities.

Before moving forward, a moment for introspection is in order. Ask yourself the following questions to ensure you have covered all aspects of this extremely important first step.

1. Have all of the skills required for all phases of the project been identified? (Refer to Figure 7-5.)
2. Have the appropriate personnel, possessing the required skills, been identified and committed to the project?
3. Has their personal commitment to the project been received?
4. Have all interpersonal skills the team members must possess to ensure the required chemistry been identified and incorporated into the selection process?

| Functional Skill Sets | Required | Not Required |
|---|---|---|
| Information technology | | |
| Finance | | |
| Accounting | | |
| Administration | | |
| Materials | | |
| Purchasing | | |
| Design engineering | | |
| Process engineering | | |
| Industrial engineering | | |
| Manufacturing engineering | | |
| Operations management | | |
| Supplier quality engineering | | |
| Quality assurance | | |
| Process quality assurance | | |
| Sales | | |
| Marketing | | |
| Field service | | |
| Order entry | | |
| Logistics | | |

*Figure 7-5. Matching skill sets at the outset with requirements for the project will pay off handsomely when the project is ramped up.*

5. Have all necessary training requirements been identified and scheduled?
6. Have all dates and locations for the project team members been identified?
7. Have the durations of the project and all subset activities been calculated?
8. Has a location been confirmed where the project team members will regularly meet?
9. Is collocation a requirement?
10. Has management's commitment to the project been received, and is their commitment wholehearted?

If the answer to each of these questions is "yes," then move forward to the next step in the process. If any of the questions were answered "no," additional groundwork must be done before moving forward.

| Project Resource Plan | | | | |
|---|---|---|---|---|
| Activity | Required Skills | Dates Required | Location(s) Required | Employee/ Candidate |
| 1 | | | | |
| 2 | | | | |
| 3 | | | | |
| 4 | | | | |
| 5 | | | | |
| 6 | | | | |
| 7 | | | | |
| 8 | | | | |
| 9 | | | | |
| 10 | | | | |

Figure 7-6. A simple matrix can match required skill sets with those who possess them.

| Training Requirements | Team Members | | | |
|---|---|---|---|---|
| Techniques | M.T. | M.G. | M.D. | S.E. |
| TQM | | | | |
| Re-engineering | | | | |
| Cycle time management | | | | |
| Statistical Process Control | | | | |
| QFD | | | | |
| Kanban | | | | |
| Supply-base management | | | | |
| Concurrent engineering | | | | |
| Agile manufacturing | | | | |
| ISO 9000/14000 series | | | | |

Figure 7-7. Training needs for project personnel also can be determined by a matrix.

## The Limits of Autonomy

Many project teams become essentially self-managed, in that the project team members make all or most of the decisions governing their actions based upon the rules and guidelines mutually established between the project team and management. Responsibility and accountability in such instances rest with the project team, making them fully autonomous. Autonomy, however, should not always be construed as complete independence. Typically, autonomy is restricted to the ability to formulate or suggest project goals, performance metrics, timing, and approach. Senior management, in most cases, defines the structure and expectations for the project, including the overall rules and guidelines under which the project team must operate. Always be cautious not to overstep your level of authority. Such mistakes tend to become career limiting.

## Team Growth

In the beginning, your new project team will not be a team at all. Rather, it will be a grouping of people with individual objectives, agendas, and approaches. They will in reality be only a committee, no matter how well you have selected them. Teams, on the other hand, are distinctively different. They must be built. If you are successful in forging your people into a team (and it can take many months to do so), the transition will become evident. Here are a few of the signs that your committee has matured into a team:

- Committees have strong, clearly recognizable leaders. Teams have shared leadership roles that change from time to time as the project progresses.
- Committees have individual responsibilities and accountabilities. Teams have both individual and shared accountability.
- Committees have individually-focused project goals. Teams have collective project objectives and expectations.
- Committee meetings are structured and rigid. Team meetings are structured, but flexible in that they focus on brainstorming and problem solving through creativity and open-forum discussions.
- Committees discuss, resolve, then assign responsibility for action to third parties. Teams, on the other hand, take responsibility for their own actions showing their commitment to one another and the team as a whole.

The transition from committee to team requires commitment from the project manager, and the individual team members. Hidden agendas must be flushed out and resolved; the team must adopt a common purpose, common approach, and common objectives; and the orientation must move away from individual to team. The team must adopt and accept shared risks and shared rewards as the basis of its charter. And, the team must develop an identity as a rallying point. As the team encounters diversity and difficulty, it will begin to unite and galvanize. Generally, the tougher the assignment, the stronger and quicker the team members will bond into a real team. But don't expect miracles overnight. The transition will take time and patience to nurture it through several predictably rough phases of development.

## MANAGEMENT VERSUS FACILITATION

So what role does the project manager take during this transition period? Success comes from being accepted as one of the team, as well as being its official leader. That often requires a completely different set of skills than the traditional project management approach. It requires facilitation skills—taking the role of aiding rather than directing in an effort to help the project team achieve its desired results. Facilitation requires the project manager to recognize how his or her personal behavior impacts the project team's performance and outlook. If the project manager is negative for example, the project team will likely see it as an indication of doubt or frustration. If the project manager is optimistic, on the other hand, the project management team will in turn be aggressive and bold in their approach to achieving the desired project objectives. Remember, the team looks to its project manager for both managerial and emotional guidance. So lead by example.

As a facilitator, the project manager must maintain open communications with the project team, and an open mind to team members' input and suggestions. Rather than giving specific assignments like a sergeant, the facilitator makes suggestions regarding the actions that are required and encourages individuals to take on each assignment. As a facilitator, the project manager must constantly monitor the project team's progress and make adjustments and corrections as necessary to guide the team back on course. The project manager must also foster confidence in his or her abilities to support the project team's needs by providing the tools and resources necessary to complete each project assignment on schedule. As goals and objectives are met, the team (not individuals) should be rewarded and encouraged to continue. And throughout the project, the project manager should recognize and acknowledge the effort being put forth by the team.

### Make Guidelines and Expectations Clear

Successfully facilitating a project team requires that decisions be made promptly in support of the team's requests and recommendations. Indecision is deadly. To facilitate the decision-making process, the project manager should always clearly define the criteria required for a decision to be made and hold the team members to those guidelines. Also, be sure that the direction, scope, and expectations for the project are clearly understood by each of the project team members. By so doing, you can rest assured that everyone is pulling in the same direction, thus making routine decisions less controversial.

Facilitation also means addressing interactions between and among team members that impact team cohesion and performance. For example, team members with strong personalities or a high degree of technical expertise can at times become overbearing, dictatorial, and loud. Such actions must be quickly addressed through one-on-one sidebar discussions between the offending team member and the facilitator/project manager, reminding the offender that everyone on the team has an equal say and that all ideas are of value to the team in achieving their objectives.

As the team becomes more comfortable with each other, there is often a tendency to accept opinions as fact rather than challenge their authenticity. In such cases, the project manager must remind the team of the need to validate all data sources to ensure complete objectivity in their conclusions and ultimate recommendations.

### Maintain Momentum

At times, the project team may begin to flounder or lose direction or their sense of urgency. The project manager must quickly reestablish the project milestones and metrics, then refocus the team's attentions on the ultimate objectives and expected deliverables. And then there is the inevitable confrontation that results from a lack of personal chemistry between the team members, or from internal politics and associated hidden agendas. In those cases, the project manager must take a stronger role than in the previously mentioned situations by quickly bringing forth the issues to the team and asking for their support to resolve the problems. If problems continue, a change of players is often the only solution. While dramatic, such actions are normally effective in resolving that and all future problems of a similar nature.

## BUILDING CONSENSUS

"Impossible!" "More trouble than it's worth!" "Too many strong or diverse opinions!" "Let's just go with a simple majority vote and get on with the program!" You've probably heard them all and more. Manufacturing engineers are generally results-oriented people who by nature believe that if you want a job done right, you do it yourself. But as project managers, you can't always take that road. You must work through the team to achieve results. That means keeping *all* team members functioning as a team, supporting the direction the team is taking and the solutions the team is proposing. And that, quite frankly, requires consensus.

But before you can reach consensus as a team, the members must all agree on what consensus means. Start with a simple definition: *Consensus means agreement on an approach or course of action that, while possibly not perfect in the minds of every team member, is an acceptable compromise that everyone can live with and agrees to support.* In essence, it is the best available alternative given all things that are considered.

How long does it take to reach consensus? It depends on many things:

• The nature and complexity of the issue before the team,
• The clarity of the goal or objective,
• The urgency in reaching a conclusion or solving a problem, or
• The mood or cooperativeness of the individual team members.

Normally, reaching consensus does take more time than a simple majority vote. But by striving for consensus, the project manager is assured that all team members have accepted the approach and that the members will continue to work together as a team for the common good.

Building consensus requires following a methodical approach that begins with agreement on what consensus means, followed by the establishment of a time limit for making the decision. Time is a great motivator in that it forces attention on the key issues while creating an environment of openness to new ideas and differing opinions. As the discussions (debates?) progress, the project manager must frequently gage the overall level of agreement between the team members and focus discussion on only those issues wherein agreement has not been reached. Do not allow the team to constantly rehash issues on which consensus has been reached, or consensus may be lost. Establish a rule early: *If a team member disagrees with a decision or approach, he or she must explain why and offer an alternative.*

If a stalemate develops, the project manager must make the final call. Before doing so, gather as much data as is possible, within a reasonable time frame. Be sure to validate all sources of data and information, then listen openly and objectively to all the alternatives being proposed and the reasoning behind each. Employ fundamental problem-solving and decision-making techniques to highlight the best possible alternative. At that time, make your decision and advise the team of your solution, and how you arrived at it. Ask for their support, then move aggressively forward to implement your decision.

## SELECTION AND APPLICATION
## OF PROJECT MANAGEMENT TOOLS

It's fact: Less than one third of all projects launched by companies today are effectively managed. And when projects fall behind schedule or exceed their budgets, project managers often look to software as a project panacea. It's true that much of the project management software on the market today is excellent for collecting, then manipulating, data into usable information. But the project manager must still provide the essential management and facilitation skills and effectively develop and execute the project plan. Decisions must be made and problems solved. These are the things that software simply cannot do. Software is simply a tool for project management.

### Problem Assessment

Let's take a look at how a project manager assesses a given situation to determine if a problem exists, or if one has the potential of developing in the near future. Situational assessments begin with a search for *change*—in organizational, functional, or individual performance; in market dynamics; in attitudes or morale; in customer service; in quality; in profits; in cycle times—in essence, a change in any of the critical elements of the project or business. The change may have occurred in the past. Or there may be indications of an impending change in the near or foreseeable future. Either way, the project manager must recognize, then address, the ramifications the change has on the project.

The project manager must begin the assessment with a series of probing questions designed to highlight what has changed, when it changed, how much it has changed, and where the change occurred. Armed with that information, he or she can begin to probe into why the change occurred and what can be done to isolate its root cause so that an effective plan of corrective action can be developed and implemented. The project manager must also look into the past to determine what problems have been left unresolved that may have a significant impact on the change identified in the present. In addition, the project manager must explore what actions have been taken in the past to resolve the problem or address the identified change, along with the reason(s) those attempts failed. The project manager must then turn to the present to isolate any actions currently under way to address the issue so as to not duplicate efforts or inadvertently hamper progress.

Once the change has been identified, all related issues or opportunities must be broken down into manageable components. Remember the old adage: "You eat an elephant one bite at a time." It's the same with problems and opportunities. They must be broken into their smallest elements so that a more thorough understanding of each element can be developed and the common links between the elements identified. One of the first questions to ask must be, "Is there more than one problem impacting the performance in several areas, or a single problem that is manifesting itself in different areas at different times in different ways?"

And because resources are typically tight, it is necessary for the project manager to prioritize which issues to pursue first based on each issue's potential for performance improvements or financial return, its level of risk or resource consumption, or the potential impact each has on completing the project on schedule and within budget.

The next step is to apply fundamental problem-solving techniques to identify the root cause that prompted the change, then employ fundamental decision-making techniques to identify the best alternative to resolve the problem.

Your initial assessment of the situation will be based on certain assumptions. Before moving forward, it is always wise to validate those assumptions to ensure you are on sound footing and that personal or organizational biases have not clouded your judgment. Remember, the validity of those initial assumptions will impact the ultimate success, costs, and timing of the project (as well as individual careers). One way to confirm your assumptions is through a simple analysis, as detailed in Table 7-2.

### Table 7-2. Validating Assumptions

| Assumptions | Level of Risk | Sources of Data | Confirmation Assigned to: |
|---|---|---|---|
| 1. Systems are adequate. | Medium | IT Manager | Bill Thomas |
| 2. System support is adequate. | High | Mfg Users | Susan Edwards |
| 3. Users are comfortable with the system. | High | Mfg Users | Jan McIntyre |
| 4. The new applications software can be installed in six months. | Medium | Mfr's Rep. | Jack Meyers |
| 5. User training can be finished by in-house personnel. | Low | Training Dept. | Shirley Shanks |
| 6. Data base conversion can be accomplished in 3-4 months. | High | IT Manager | Bill Thomas |

### Problem Solving

The basis for an accurate evaluation of a given problem is cause-and-effect reasoning: isolating what has gone wrong or what has caused the identified change to occur. A problem is the visible effect of a cause that has occurred at some time in the past. The key is to relate that exact cause to the effect you are observing, because only then can you effectively resolve the problem and keep it from recurring. Problem solving begins with a concise, accurate definition of the problem derived by answering the following questions:

1. Exactly what problem am I trying to resolve?

2. Where was the problem first observed?
3. When did the problem first appear?
4. What is the significance or magnitude of the problem?

In defining the problem, it is essential that the project team isolate both what *is* and what *is not* happening. This approach provides the basis for comparing what actually is happening to what should be happening, and gives an indication of the size of the gap between the two. It allows the project team to then identify each particular factor separating the existing situation from the desired conditions.

As before, the project team should look for changes that have occurred in each of the should-be conditions. Once identified, the team should explore the what, where, when, and how much of each. The true root cause of the problem must answer each of the four questions. Remember, only the true root cause of the problem can create the effect you are observing. No other cause can. Therefore, once the root cause has been identified, it must be confirmed through a re-creation of the problem. Only then can the project team be sure they have found the basis of the problem.

Commonly used problem-solving tools include:
- Evaluation and planning tools (brainstorming, flow charts, cause-and-effect diagrams);
- Data collection tools (surveys, focus groups, questionnaires, interviews, measurements, and observations); and
- Data display tools (run charts, histograms, check sheets, Pareto charts).

The use and application of these tools is widely understood and practiced by most manufacturing engineers in their work and, thus, will not be discussed further at this point. However, additional discussion of data collection methods is in order, as often the data used during problem-solving sessions is found to contain a significant amount of error. There are several rules to follow in collecting data to ensure that additional variables are not introduced into the evaluation process.

When collecting data, the first step is to identify exactly what it is you want to know or learn, and then to determine both the type and amount of data necessary to support your problem-solving efforts. Collect the data, then sanitize it by correcting obvious errors and filling in gaps. Be cautious. Do not assume the data is accurate as collected, even if it is computer generated. Inaccurate data can be misleading, allowing the project team to reach the wrong conclusions and initiate invalid corrective actions. Next, organize the data by type, category, time, frequency, etc. Then condense it by computing descriptive statistics such as averages, mean, or range. And, finally, convert the data into information using the data display tools identified earlier.

Always document your work: the assumptions the team has made; the data collection methods; the data conversion methods and their results; the steps taken by the team to validate their data; the conclusions reached from the team's assessment of the data; and the team's corrective action plans and the results achieved from them against project expectations.

Once the root cause of the problem has been identified, a decision must be made regarding the best alternative to resolve the problem.

### Decision Making

Many project managers avoid making decisions because of the controversy and direct confrontations commonly associated with the decision-making process. Often, the project manager's decision results in a contest between differing points of view. Politics often come into play, in which case, the person with the most political clout prevails and the loser suffers the embarrassment of defeat in front of his or her project team, peers, and superiors. For these reasons many project managers avoid making decisions.

But as a project manager, one of your responsibilities is to make decisions, sometimes unpopular ones. In short, decision making is critical to success; indecision is deadly to your project team, their efforts, and your credibility.

#### What's Involved in a "Good" Decision?

Any good decision is based on a comprehensive understanding of the project requirements, coupled with the results of a thorough problem-solving effort which has isolated the root cause of the problem impacting project performance, cycle time, resource consumption, costs, etc. With this information, you can make a complete assessment of the alternative courses of corrective action available, followed by an analysis of the results you can expect from each of the alternative solutions—good and bad.

There is no pressure on you to make a "perfect decision," because there simply is no perfect decision nor ideal solution. The best decision is one that results in the solution coming closest to fulfilling all project requirements and expectations in the most cost- and time-effective manner possible, with the least amount of risk. And remember, current conditions are an alternative that must be considered.

The decision process begins with an accurate description of what the decision is intended to address or resolve (the decision objectives), and the known alternatives available to the project team to accomplish those objectives. The objectives established for the decision process are listed, then divided into two categories; those that *must* be achieved to guarantee the success of the project, and those the project team would *like* to achieve, but are not necessarily mandatory to the successful completion of the project. "Like" objectives are those that are used for comparison of one alternative against another. They are the differentiators in the decision process.

All "must" objectives are quantifiable and measurable. They function as go/no-go qualifiers for all alternatives. Unless an alternative can fulfill all "must" objectives, it cannot meet all of the factors that are essential for a successful decision and, thus, the successful completion of the project. Therefore, any alternative that does not satisfy all "must" objectives is immediately removed from consideration.

Every decision has consequences, both positive and negative. Those consequences must be considered and assessed for their potential impact on the project and the risks they introduce relative to the successful completion of all project objectives. And they must be considered *before* a final decision is reached. It is the only opportunity the project team will have to address any negative consequences at little or no cost to the project. Negative consequences will always introduce additional problems. As Shakespeare said, "The evil that men do lives after them, the good is oft interred with their bones." The same can be said for a poor decision. It will outlive all of the good work done by the team and the project manager.

*Overlooking negative consequences that make a decision unworkable or project objectives unreachable, is a fundamental and potentially career-limiting mistake that no project manager wants to make.*

*Prioritizing Decisions*

Every decision objective must be prioritized according to its relative importance to the project requirements and expectations, as well as for comparison against all other decision objectives. In addition, the prioritization of project objectives will force the project manager to evaluate the effectiveness of the project team in establishing the priorities. For example, if a large number of objectives carry a high priority, the expectations may well have been set too high; and vice versa.

One of the best aids in structuring the decision process is a matrix, similar to the one in Figure 7-8, which contains all the "must" and "like" objectives, along with a prioritization scale or weighting factor which differentiates one objective from the others based on their relative importance in meeting the overall project requirements. But remember what we covered earlier: a "must" objective carries no prioritization. It is a requirement that, unless met fully, completely eliminates the alternative under consideration.

| Project Management Decision Objectives Matrix | | | | | | |
|---|---|---|---|---|---|---|
| **Must Objectives** | **Alternative A** | | **Alternative B** | | **Alternative C** | |
| A (Description) | Yes | No | Yes | No | Yes | No |
| B | | | | | | |
| C | | | | | | |
| D | | | | | | |
| **Like Objectives** | **Wt** | **Score** | **Wt** | **Score** | **Wt** | **Score** |
| A (Description) | | | | | | |
| B | | | | | | |
| C | | | | | | |
| D | | | | | | |
| E | | | | | | |
| F | | | | | | |
| **Total Scores** | | | | | | |

*Figure 7-8. Categorizing objectives simplifies the project manager's decision-making.*

"Like" (or "desired") objectives, are weighted on a 10-point scale, with a weighting of 10 indicating the highest level of importance in meeting project or decision requirements, down to a 1 indicat-

ing the least significance to the overall project or decision requirements. Each alternative is then scored against each objective using the following scale:

- 5 points–100% conformance in meeting the "like" objective,
- 4 points–80-99% conformance,
- 3 points–60-79% conformance,
- 2 points–40-59% conformance,
- 1 point–20-39% conformance, and
- 0 points–less than 20% conformance.

Scores for each alternative are calculated by multiplying the point
totals assigned by the weighting factor for each "like" objective,
then totalling all the scores for that alternative. The alternative
that best conforms to the decision or project objectives, based on
the highest score, is the recommended course of action or solution
to the problem.

A word of caution, though. Apply a reality check to your proposed solution or corrective action. Is it workable within your environment? Is it implementable within your organization in a
reasonable time frame? Is it economically feasible? Have all functional, organizational, and political barriers been breached? Have
all necessary management approvals been received? If so, move
forward to implement your decision. If not, further review and additional actions are necessary.

Now that we have discussed the fundamental situational assessment, problem-solving, and decision-making tools used by project
managers, let's take a look at some of the project planning tools
available.

### Project Management Planning Tools

#### Work Breakdown Structure
One of the most common tools used by project managers in the
early planning stages of a project is the *work breakdown structure*
(WBS). The WBS, much like a bill of material, breaks the project
down into the major sublevel or subtier activities required to complete each parent activity for the purpose of providing an accurate
overview of the entire project—including all of the major activities
that must be completed to achieve the project objectives (see Figure 7-9). This, in subsequent planning stages, provides the basis

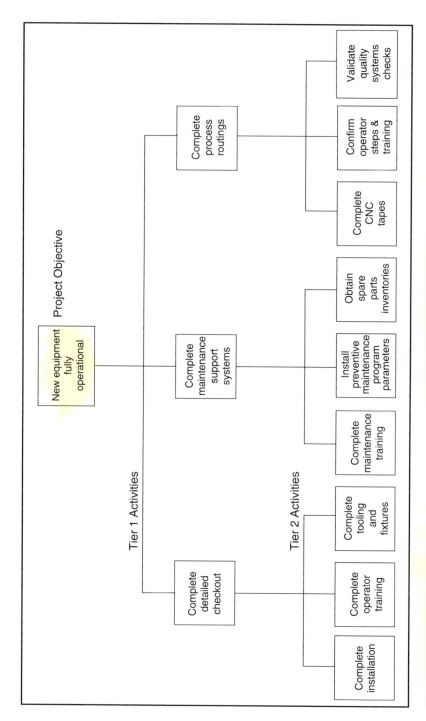

Figure 7-9. A work breakdown structure identifies tasks, establishes dependencies, and aids in setting priorities.

for the creation of the project's activity listings and all associated interdependencies.

The first step in the creation of the WBS is the definition of the project objective. Then, all major activities or categories of activities that are required to accomplish the project objective are defined. These become the Tier-1 activities—the major milestones of the project. Next, the Tier-2 activities are defined. Just as the Tier-1 activities represent the child-parent relationship with the project objective, the Tier-2 activities represent the child-parent interdependencies with the Tier-1 activities. Thereafter, each subsequent subtier of activities represents the requirements for completion of the next higher tier of activities until sufficient detail is available with which to accurately describe the project in the aggregate. Then, like a bill-of-material cost rollup, the total project cycle time and project costs can be forecasted.

Use of the WBS in the early planning stages of the project prevents the omission of critical project activities because it pictorially displays each of the major activities that must be addressed to complete the next higher tier activity, along with all associated interdependencies. It is much like the creation of an outline prior to writing a paper or report. It defines the scope and content of the project for subsequent refinement.

### Activity Listing

The next step in the project planning cycle is the creation of a detailed *Activity Listing* for each tier of the WBS. This is where the meat is added to the project skeleton. For each tier of the WBS, a separate activity listing should be created to provide the necessary level of detail to ensure that all critical issues are adequately addressed. Each activity should be described in enough detail to ensure clarity; it should then be assigned a unique number so it can be monitored against the project schedule. The duration of each activity should be calculated or estimated, along with its expected start or completion dates, as illustrated in Figure 7-10.

The Tier-1 activities in Figure 7-10 are, in effect, the primary project milestones—benchmarks for the project signifying the accomplishment of key project deliverables. Milestones for Tier-1 activities are assigned in Tier-2; for Tier-2 in Tier-3; and so forth. Milestones are typically used for planning and evaluation purposes

| Task Number | Description | Estimated | | Actual | |
|---|---|---|---|---|---|
| | | Time | Completion | Time | Completion |
| Tier 1 | | | | | |
| 1A1. | Draft rookies | 90d | 2/96 | 85d | 1/96 |
| 1A2. | Sign veterans | 180d | 6/96 | 270d | 10/96 |
| 1A3. | Sign free agents | 60d | 8/96 | 60d | 8/96 |
| Project Description: Win the Super Bowl | | | | | |
| Project Manager: Jimmy Johnson | | | | | |
| Project Team: Your guess is as good as mine. | | | | | |

*Figure 7-10. Further detailing the work breakdown structure, the activities list ensures that all project issues are properly dealt with.*

throughout the project as a means of monitoring performance against established targets.

### Precedence Diagramming

Another consideration for the project manager is interdependencies. The dependencies between activities can be critical, and thus require careful consideration during the project planning cycle. And dependencies can be either time or resource sensitive. One project management tool especially useful in highlighting dependencies is the *Precedence Diagramming Method*, or PDM. An example of which is shown in Figure 7-11. The PDM is used by project managers to illustrate the precedence and interdependencies of activities that relate to one another. As illustrated in the example shown next, the first activity that must be completed is shown at the far left of the precedence diagram. Each subsequent activity is then shown to the right in the sequence of its planned occurrence, based upon either its time or resource dependency. The different paths within the diagram illustrate the activity flows within the project, the longest of which becomes the critical path.

### Cycle-time Estimating

The next task is the estimation of the *cycle time* for each activity in the project, and the subsequent rollup of the cycle times of all activities along the critical path to determine the total project cycle time. As in our re-engineering activities, cycle time is the total

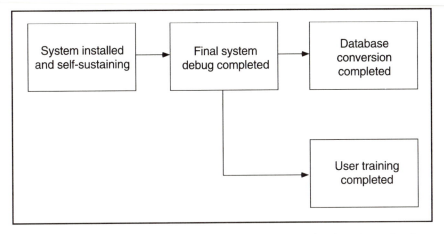

*Figure 7-11. Precedence diagramming clearly illustrates project interdependencies.*

time consumed from the beginning of an activity to its completion, including all value- and nonvalue-added times. Several issues deserve consideration when estimating the cycle time of an activity. First, the complexity of the activity can add greatly to the time required to complete it. Simpler activities can be completed quicker. A good rule of thumb is to utilize standard industrial engineering time-and-motion studies to assess the actual cycle time for the same or similar activities.

Another tool can be taken from our setup reduction discussions earlier in this book—the videotape methodology. And, as in our re-engineering review, the owner of the activity should be consulted to ensure all aspects of the activity have been included in the cycle-time assessment.

The productivity level of the people performing or projected to perform the activity should not be assumed to be 100%. Where available, factor actual productivity measurements into your cycle-time calculations. Where productivity figures are not available, a conservative estimate of 50-55% should be used.

Another factor in the cycle-time calculation is the degree of concurrency that can be employed, as evidenced in the precedence diagram. The more activities that can be done in parallel, the shorter the cycle time.

The accuracy of the activity description, work flow, or process specifications will add greatly to the ability to perform the activity in the shortest possible cycle time. Conversely, the lower the level of accuracy, the longer the activity will take to complete.

An alternative in calculating the cycle time for each activity is the *Program Evaluation and Review Technique* (PERT) mathematical model:

$$CT = \frac{(OT) + (4 \times MLT) + (PT)}{6}$$

where:  CT  =  Cycle Time
OT  =  Optimistic Time
MLT =  Most Likely Time
PT   =  Pessimistic Time.

Whichever methodology you employ, be as conservative and as accurate as possible within the constraints of the project.

*Calculating Calendar Time*
It is now necessary for the project management team to accumulate the individual activity cycle times and convert them into actual calendar time. In so doing, remember to consider scheduled down periods and weekends. If unscheduled downtimes are common due to equipment breakdowns, supplier delivery problems, or any similar factor, include in your calculations a realistic estimate of the projected lost time. Also consider:
- Absenteeism,
- Training requirements,
- Available resources,
- Organizational priorities,
- Learning curves,
- Availability of critical equipment,
- Administrative cycle and approval times,
- Labor contract expirations,
- Customs or international processing times,
- International holidays,
- Customer and supplier issues,
- Turnover,
- Employee skill sets,
- Functional priorities,
- Work backlogs, and
- Queue times.

*Charting Your Progress*

At this stage, most project managers employ PERT through computer models to calculate both the critical path and the slack times within the project schedule. The critical path, as mentioned, is the shortest series of activities within the project plan, found by summing all of the individual activity cycle times in every conceivable path throughout the network of activities to determine the shortest path. The critical path is dynamic in that it is dependent on the cycle times of the activities which collectively comprise it. As they slip or compress, the critical path is likely to be impacted accordingly. And as the critical path goes, so goes the project schedule.

Every path through the network of activities, except the critical path, contains some degree of slack or dead time. By identifying the slack time, where it exists and to what degree, the project manager can better manage the project by reallocating limited resources from noncritical path activities to critical path activities. This will compress the project schedule if an unavoidable slip in one or more of the critical path activities occurs.

The final and most widely used of the project management tools is the *Gantt chart*. The Gantt chart illustrates the time lines of all project activities, as well as the total project cycle time. In addition, it highlights opportunities for the project manager to overlap activities or run them in parallel to further compress the project schedule. Interdependencies are evident on the Gantt chart, as are resource constraints. Milestones and critical path activities are also illustrated. In short, the Gantt chart becomes the compilation of all the project management planning tools; thus, its value and popularity to project managers.

But Gantt charts, if constructed properly using the tools and techniques mentioned, require many hours of painstaking work. To facilitate this arduous task, and increase the accuracy of the work at the same time, most project managers employ computer models at this stage in the planning process. So let's take a brief look at what the computer *will* and *will not* do for you.

**Project management software.** With most project management software on the market today, data entered by the user (activities, interdependencies, dates, durations, resources, etc.) is converted into Gantt charts or PERT charts, with the critical path calculated and identified. User friendliness and flexibility are the two key elements differentiating most software products. Costs are dependent upon the platform to be used and the functionality required by the user, typically ranging from less than $100 to over $5000.

Considerations in selecting the right software for your project are numerous, but not unmanageable. First, there are the obvious:

- On what type of platform (equipment) will the software be operated? In what type of environment (standalone, file server, client server, network, etc.)? Using what operating system?
- How much memory is required in both RAM and ROM to support the software and ensure its responsiveness?
- What type(s) of printer(s) does the software support? Tape drives? CD-ROMs?

In sizing your software, give consideration to the number of activities your project includes, as well as the increments of time you want to use (day, week, month, year, etc.).

Also consider the capabilities and features that will make the software you select work effectively for you:

- Graphical user interface (GUI);
- Concurrent, as well as serial activity capabilities;
- Availability of or compatibility with the basic project planning tools discussed earlier in this chapter (WBS, activity listings, PERT, critical path, milestones, Gantt, etc.);
- Parent-child dependency links;
- Project cost management capabilities (progress payments, budgeting, multiple billing rates, multiple currency rates, etc.);
- Ease in removing and adding activities;
- Resource tracking and dependency capabilities;
- Resource leveling capabilities;
- Tie-ins to other application software programs;
- On-line help screens; and
- On-screen tutorials.

Other considerations include technical support lines, reputation, user groups, and installed sites. In general, most project management software products on the market today are sound, proven products. The key is to select one that fits the size of your project and is compatible with the hardware you are using. Only in rare circumstances is it necessary to develop your own custom programs. With the capabilities designed into the current off-the-shelf products, resist the temptation and the risk.

## Resource Leveling

During most projects, there comes a time when either resources or budgetary constraints require the project manager to realign the project schedule or spending to balance available resources with either business or organizational requirements. *Resource leveling* (Figure 7-12) is a technique used by experienced project managers to spread project costs evenly or to allocate limited resources (like technical support or skilled labor) to critical-path activities using the available slack time in the project schedule. Resource leveling also provides the project manager with insight into the rate of consumption of those limited resources so that adjustments can be planned in advance, not implemented in reaction to a slip in the project schedule or an over-spent condition.

In essence, resource leveling involves a reallocation or shifting of project activities to comply with available project funding or other limited resources. For example, tight or unexpected financial constraints on the organization could force an extension of the

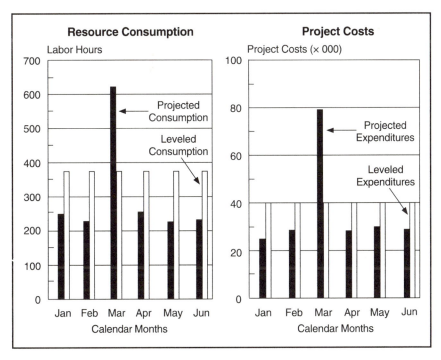

*Figure 7-12. A proven method for allocating manpower resources and spreading project costs is resource leveling.*

project schedule in order to balance the project budget, or hold project costs to a fixed level of spending per time period. In other cases, a similar technique—*resource shifting*—is used to draw limited resources from noncritical path activities to critical path activities to hold or compress the critical path cycle time by consuming the slack time within the project schedule.

The key is foresight. By effectively managing the project and monitoring its progress against established project metrics, the project manager can effectively utilize these techniques to control project costs and timing. Again, *prevention is always less costly than corrective actions once a problem has occurred.*

## PROJECT COSTS AND JUSTIFICATION

Every investment a business makes must produce an acceptable return. The same is true for the project. It must produce tangible, measurable financial results or performance gains. Otherwise, it simply should not be undertaken. It is, therefore, necessary that the project manager carefully calculate the project budget, monitor spending to stay within those budgetary constraints, and ultimately determine the return on investment (ROI) that has resulted from the project.

Let's start with the establishment of the project budget. True, it is normally a dictated number, but shouldn't you decide early in the game whether or not you have a chance of meeting your financial constraints?

Project costs are generally broken down into several categories such as those in Table 7-3. Any one or all of these categories may apply to your project.

Tracking project costs in the aggregate and by individual activity gives the project manager a high degree of fiscal control over the project. Managing a project is like managing a small business. There is only so much money allocated, and thus only that amount can be spent. There is no going back time and time again for more allowance. You are a manager charged with controlling every aspect of the project, including its budget. If one activity goes over budget, another must come in under budget to compensate for the overage. Funds must be managed over time. They cannot all be spent within the first few days of the project. Rather, they must be

## Table 7-3. Project Cost Elements

*Project Management Costs*
Personnel, travel and expenses, support systems, training, software, hardware, supplies, miscellaneous expenses, etc.

*Operating Costs*
Utilities, telephone, facsimile, copies, printing, occupancy/rental, maintenance supplies, perishable tooling or test equipment, outside services, local transportation, samples, etc.

*Capital Costs*
Capital equipment, finance costs, spare parts, operation and maintenance training, freight, duties, royalties, customs fees, taxes, installation and/or erection costs, R&D costs, support systems costs.

*Project Costs*
Material, labor, burden, freight, packaging, handling, overhead.

controlled to ensure there is sufficient capital to support the project throughout its duration.

The matrix shown in Figure 7-13 is a useful tool for monitoring project expenditures. Much like a checkbook, this simple project cost matrix tracks expenditures and ending balances for each activity and for the project as a whole. Costs should be tracked daily, with expenditures stringently controlled, and contingency plans

| Cost Elements | Activity | Date | Activity | Date | Activity | Date |
|---|---|---|---|---|---|---|
| Beginning balance | | | | | | |
| Labor | | | | | | |
| Burden | | | | | | |
| Material | | | | | | |
| Overhead | | | | | | |
| Capital | | | | | | |
| Operating costs | | | | | | |
| Project management costs | | | | | | |
| Ending balance | | | | | | |
| Project balance | | | | | | |

*Figure 7-13. Budget visibility is facilitated by a project cost matrix which tracks expenditures and ending balances by project activity.*

implemented before an over-budget condition occurs. Otherwise, there may be no way of recovering.

Another word of caution: be alert for arbitrary allocations of overhead and burden to your project. Unless a given department, function, process, or another project is impacted by the results of your project, its overhead and burden should not be applied to *your* project budget. Every activity and every project must stand on its own. Accountants are not dishonest. They simply look for the easiest way to distribute hard-to-allocate costs. The path of least resistance usually ends up with the lion's share of the unallocated imbalances. It is therefore necessary that you monitor your project costs carefully and frequently.

As was noted, the objective of most projects (and all senior managers) is to maximize return on investment for the organization. The calculation of ROI has a number of variations ranging from the simple ratio of project savings to project costs, to the more complex calculation of return-on-assets times financial leverage. The method you select to calculate the ROI for your project must be compatible with that used by your particular organization. Thus, it is recommended that you solicit the assistance of a financial analyst or accountant when calculating your ROI. These people know how it's done *and* they have credibility when it comes to numbers. They also have an uncanny knack for uncovering costs or savings overlooked by the project team. Remember, there are few things more embarrassing than a "surprise" during your presentation to senior management. So take the safe approach.

## REVIEW AND REFLECT

Throughout the project, the project manager should frequently reassess the project objectives and status to ensure the project remains focused correctly. Changes in conditions, assumptions, or scope could lead to completion of the project without achievement of the expected deliverables. Outside influences can occur which make the original project objectives unachievable. Inadequate or untimely allocation of critical resources can push the project schedule beyond the expected completion date. Ineffective cost controls can cause budget overruns. In short, numerous internal and external factors can occur which negatively impact project performance, timing, or cost. It is therefore incumbent upon the project manager to constantly monitor every possible variable to ensure continued compliance to the project plan. One way of so doing is to maintain a running project checklist like that shown in Figure 7-14. Constant critical assessments, like this, will go far in guaranteeing that no surprises arise to trip you.

## ASSESSING PROJECT ALTERNATIVES

Sometimes, even with the best-managed projects, an unforeseen problem arises that impacts the expected project deliverables, throws the project off schedule, or creates an over-budget condition. In such cases, the project manager must immediately assess the alternatives available to bring the project back under control. The *focus* of those alternatives *must always be the customer.*

### Internal Issues

The project manager should first explore the internal issues like: "Can the project's scope be redefined without compromising performance, cost, or schedule?" Thereafter, other internal issues should be explored. For instance:

*Project Schedule*
- By reallocating resources from noncritical path activities to critical path activities, can the project timing, cost, and/or performance be brought back into expected ranges?
- Can any of the noncritical activities be eliminated or reprioritized to free critical resources?
- By applying additional resources, can the planned project completion date be achieved?
- Have we employed all compression techniques like concurrency and resource leveling to reduce the total project cycle time as much as possible? How about overtime?
- Can we assign noncritical project activities to administrative or clerical personnel to free project team members with the needed skills to take on other more critical project assignments?
- Have we considered subcontracting project activities that can be effectively controlled on the outside, or that offer a minimum risk of failure or delay?

| Project Management Checklist | Yes | No |
|---|---|---|
| Have all activities been identified and listed? | | |
| Has the cycle time of each activity been calculated? | | |
| Have the necessary team skills been identified? | | |
| Have the correct team members been selected? | | |
| Is the team working well as a unit? | | |
| Are assignments being completed on schedule? | | |
| Is the team's problem-solving approach effective? | | |
| Are critical decisions being made on a timely basis? | | |
| Are any hidden agendas arising? | | |
| Is the team maintaining focus? | | |
| Has the team's time been properly allocated? | | |
| Has the critical path been identified? | | |
| Have all project planning tools been employed? | | |
| Has all slack time been identified and reallocated properly? | | |
| Have critical milestones been identified? | | |
| Have management review schedules been set? | | |
| Have all support systems been confirmed? | | |
| Are the conditions the same as when the project was launched? | | |
| Are the initial objectives still valid? Attainable? | | |
| Have all assumptions been reviewed and validated? | | |
| Has the scope of the project changed? | | |
| Has the project been impacted by any outside influences that could negatively alter project schedules, cost, or performance? | | |
| Has management altered the original priority assigned to the project? | | |
| Have project expectations changed since the project was started? | | |

Figure 7-14. *Project status is always current if a running checklist is maintained. This type of checklist provides the project manager the flexibility to adjust and refine the scope, focus, and objectives of the project.*

- Have we re-engineered all processes effectively to:
  - Minimize their cycle times?
  - Incorporate concurrent methodologies?
  - Eliminate all waste and redundancies?
  - Highlight and capitalize on islands of opportunity?
  - Identify and address all process, material, and information bottlenecks and queues?
- Have we automated where possible to minimize cycle time?
- Have we effectively isolated and utilized all pads (slack time) in the project schedule?
- Have all unnecessary approvals been removed to further improve cycle time?
- If none of the alternatives is plausible, does the extension in the schedule provide additional benefits to the customer that will make the delay acceptable?

*Project Costs*
- If the project is over budget, what was the reason?
- Can other project costs be reduced to offset the overruns?

*Project Performance*
- Can the original performance targets be met by applying additional resources or capital?
- Can the original performance targets be met by applying other technologies or methodologies available to the project team?
- Can the original internal performance targets be reduced and still meet customer or market expectations?

**External Issues**
If all avenues have been exhausted internally, the project manager must address the external issues, typically the less pleasant ones.
- Was the customer responsible for the project delays, cost overruns, or changes in performance criteria?
- Will the customer accept the expected delays in the project schedule? Cost overruns? Performance differences?
- Can one of the three critical project criteria (timing, cost, or performance) be renegotiated to bring the other two back into compliance with expected project deliverables? For example:
  - Can the performance or schedule criteria be renegotiated to allow the project to remain within cost guidelines?

– Can project cost be compromised if the project remains on schedule and achieves expected performance targets?
– Can any combination of the above two provide the customer with the best possible outcome?
• Have we correctly identified the customer's priorities relative to schedule compliance, project cost, and expected project performance?
• In selecting our alternatives, have we adequately assessed all possible liabilities resulting from our failure to comply with the project requirements? For example:
  – Is our failure to comply with one or more of the project requirements likely to impair our ability to obtain additional business from this or any other customer?
  – Are there any financial liabilities or penalties associated with our failure to meet the project requirements?
  – Are there any downstream or contingent liabilities associated with our failure to meet the project requirements that may not be evident at this time?
  – Are we exposing our organization to product liability or personal injury litigation by our failure to comply with project performance requirements?

### Calculating Alternative Metrics

The list of alternatives and considerations can be lengthy, or rather short, depending on your business and your project customers. As the project manager, it is your responsibility to isolate as many feasible alternatives as possible, then assess each to determine which represents the best possible outcome for the customer. To do so, first list the project objectives and project alternatives as illustrated in the matrix of Table 7-4.

Then, assign a weight to each of the objectives that is consistent with its relative importance to or impact on the customer. Finally, assign a probability of success to each alternative in meeting each of the respective objectives. The numbers will lead you to the alternative that provides the highest probability of success in meeting the objectives you have established for project cost, timing, and performance.

The corrective action you select must address the root cause of the project problem. And it must be measurable in time and or

### Table 7-4. Alternative Assessment

| Objectives | Meet Cost Requirement | Meet Performance Standards | Meet Delivery | Increase Business | Maximize Margins | Total |
|---|---|---|---|---|---|---|
| Weights | 0.20 | 0.50 | 0.25 | 0.025 | 0.025 | |
| *Alternatives* | | | | | | |
| Request extension | 95% | 100% | 0% | 50% | 100% | 72.75% |
| Absorb added costs | 100% | 90% | 85% | 75% | 0% | 88.13% |
| Miss delivery | 80% | 95% | 0% | 10% | 90% | 63.5% |
| Add resources | 50% | 95% | 95% | 90% | 0% | 83.5% |
| Reduce quality | 85% | 50% | 85% | 15% | 90% | 65.88% |

dollars, as well as be implementable within a short time frame. The longer it takes to implement the corrective action you have selected, the higher the probability that project objectives will not be met or, at the least, compromised significantly. Finally, remember to choose the alternative which best addresses the customer's requirements and their respective priorities.

### Communicating and Documenting Your Efforts

Once the alternative is selected, the customer must be immediately advised of your intended course of action so that he or she can plan accordingly. Remember, like elephants, customers who receive last-minute surprises regarding project delays, cost overruns, or performance deficiencies have a tendency *never* to forget.

Next, document all of your assumptions, corrective actions, and their results clearly and comprehensively. The alternative selected to address the project performance, budget, or schedule problems should be supported by the data used in its selection, along with its associated implementation plan. This is a cumbersome task, but it is necessary. If the customer sees that you have done an exhaustive job of isolating, prioritizing, assessing, and selecting all feasible alternatives to correct the project problem, they will at least be assured that everything possible has been done to meet their requirements. Who knows, they may even think more of you for your efforts on their behalf.

## IT'S UP TO YOU

Project management can no longer be taken lightly, as was often the case in the past. It requires both a high level of commitment and effective management to be successful. So before embarking upon project management assignments, take the necessary time to reflect on what you are committing to and what you must do to make it a successful assignment. Plan effectively. Control all variables. Select the right team members. Manage time and budget with equal care. Make decisions based upon sound techniques and good data. When problems occur, address them quickly and decisively. Manage your projects as you would manage your own business.

# 8

# Managing the Supply Chain

As we touched on briefly in Chapter 2, successful manufacturers today are concentrating on building solid relationships with a few key suppliers deemed critical to their business operations, process quality, and operational overhead. To effectively manage the three key manufacturing metrics (inventory, operating expense, and manufacturing cycle time), manufacturers have employed supply-chain management techniques to quickly and quantitatively assess which suppliers are worthy of their business, and then to reduce their supply base for each critical commodity to its lowest possible level, thereby maximizing control while minimizing material and operating costs.

Successful supply chain relationships have yielded several measurable benefits:

- A controlled reduction in a manufacturer's supply base, affording increased operational control over the procurement process;
- More and longer-term business for those suppliers selected, resulting in lower initial material prices and sustained price reductions over time for the manufacturer;
- Higher quality, resulting from a stronger emphasis on meeting the manufacturer's requirements through process control versus inspection and appraisal techniques;
- Lower total costs for the manufacturer, resulting from reduced supply-chain lead-times, enhanced JIT inventory capabilities, and lower material management and handling expenses (including sorting, scrap, and rework of defective supplier materials); and finally
- Enhanced on-time delivery compliance from the supply chain which, in turn, allows the manufacturer to reduce safety stocks and manufacturing cycle times.

The initial supply-chain management techniques were directed by either the purchasing or quality functions. And while they met with some success, those initial approaches focused primarily on the suppliers' quality systems and the manufacturing equipment available to produce the manufacturer's products. Issues such as process control and process capability were given superficial treatment. As a result, only minimal improvements were realized, while risks grew as more and more business was placed with fewer and fewer suppliers.

Today's world-class supply-chain management techniques are much broader based, focusing on all supply-chain processes that contribute to or influence the suppliers' quality, delivery, count, and lead-time performance. Risks are minimized as quantitative approaches are employed to monitor all relevant elements of the supplier's operations to ensure consistent and continued conformance to all customer requirements. And because these efforts are more enterprise-focused, additional expertise is now required in the supply-chain evaluation processes—expertise embracing that provided directly by the manufacturing engineer.

Successes today are endless in supply-chain management methodologies. Manufacturers as diverse as Motorola, Asea Brown Boveri, Schering-Plough Healthcare, Brown & Williamson Tobacco, John Deere, Sara Lee Intimates, J.I. Case, Hallmark Cards, Inc., and numerous others have successfully reduced their supply chains while realizing increased material quality at lower total costs. Savings of 7–12% in annual material procurement expenses are common, amounting to millions of dollars in cash-flow improvement annually for those manufacturers who have deployed these methodologies successfully.

The single largest component in the product cost calculation for most manufacturers is material, typically amounting to more than 60% of the total. It is only logical, then, that control and management of that component should receive the lion's share of management's attention. And the manufacturing engineer's input in those processes is critical. After all, who is better suited to evaluate a supplier's manufacturing process capabilities, process controls, and process capacities. It is because of that critical expertise that successful manufacturers have enlisted their manufacturing engineers to take an active role in the selection and certification of critical suppliers.

## ASSESSING THE SUPPLY CHAIN

When evaluating a supplier, three fundamental performance metrics must be demonstrated to ensure ongoing conformance to the manufacturer's dynamic material requirements:

- *Process capability*, as measured statistically for all critical processes associated with the production of the products supplied by the producer to the customer;
- *Process control* for all operational and administrative functions and activities supplying information, materials, tooling, equipment, documentation, labor, and technology to those same production processes; and
- *Commitment to continuous compliance* with all customer requirements, every day, every shipment.

In most organizations, conformance to those performance metrics typically requires a cultural revolution. The need to develop a true customer-focused enterprise in which all elements of the producer's organization from order entry to shipping are focused on and dedicated to meeting the highest expectations of customer satisfaction attainable essentially mandates significant organizational change.

The focus of this chapter is on providing the manufacturing engineer with information on how those supply-chain tools are developed and deployed. Armed with this understanding of the creation and use of new tools, the new manufacturing engineer is better equipped to satisfy the manufacturer's requirements for reducing the supply base, material costs, and operating expenses, as well as minimizing the risk of unexpected material shortages or nonconformances.

## SETTING THE STAGE

It's appropriate to begin with a piece written in the early 1900s by an unknown author whose work still accurately describes consumers, their needs, and their expectations.

*"I am your customer. Satisfy my wants. Add personal attention and a friendly touch, and I will become a walking advertisement for your products and services. Ignore my wants, show carelessness, inattention, and poor manners, and I will simply cease to exist as far as you are concerned.*

*"I am sophisticated. Much more so than I was a few years ago. My needs are more complex. I have grown accustomed to better things. I have money to spend. I am an egotist. I am sensitive. I am proud. My ego needs the nourishment of a friendly, personal greeting from you. It is important to me that you appreciate my business. After all, when I buy your products and services, my money is feeding you.*

*"I am a perfectionist. I want the best I can get for the money I spend. When I criticize your products or services, and I will to anyone who will listen when I am dissatisfied, then take heed. The source of my discontent lies in something you or the products you sell have failed to do. Find that source and eliminate it, or you will lose my business and that of my friends as well.*

*"I am fickle. Other businessmen continually beckon to me with offers of more for my money. To keep my business, you must offer something better than they. I am your customer now, but you must prove to me again and again that I have made a wise choice in selecting you, your products, and your services above all others."*

What performance can you expect from your supply chain? Is zero defects really a reasonable expectation? What about 100% on-time deliveries? Or a true JIT methodology? In most cases, even seasoned manufacturing and quality professionals question such lofty goals. They accept performance levels that they are told are consistent with "industry standards." The thing is, I have yet to meet the experts who established those so-called standards. I always looked at it this way: My money represents 100% quality. It spends on time, every time. It is always 100% accurate in its value. Now, if you want it from me, you must give me a like-kind product in return. If the supplier wants to ship me materials with 90% quality levels, why shouldn't I pay them 90 cents instead of a dollar? Seems reasonable to me. After all, if it was *your* money versus your organization's, wouldn't you expect the same? Why then do we accept less when spending the company's money?

Well, Mike, you just don't understand.

Really?

Try this.

You just bought a new car worth $35,000. You got all 10 of the options you have been saving for. It's a beautiful car. You drive it home. As you are driving, you begin checking out each option. They all work beautifully. All, that is, except your new stereo AM/FM CD radio. Mad? Why? You got 9 out of 10 of the options you ordered to operate to your expectations. Isn't that the same 90% quality you accept from your suppliers at work? And, oh by the way, the brakes will only work 9 out of 10 times, so keep an accurate count.

Get the picture?

Maybe 100% isn't reasonable today. But why not establish stretch goals for your suppliers and assist them in meeting those performance levels? It worked for Motorola and the others. It can work for you. Then, once they meet those goals, move the target continuously until, together, you achieve six-sigma levels of quality and delivery.

Here's how the world-class manufacturers did it.

## SUPPLY-CHAIN MANAGEMENT METHODOLOGY

The secret behind effective supply-chain management is a structured methodology that utilizes the following approach to maximize control and minimize nonconformances. The approach has a number of variations that customize it to the particular manufacturer's environment and operating constraints. There is no "canned" approach that fits everyone. So my recommendation is to take those elements of this approach that fit your situation, then customize them to your own unique operating constraints and requirements.

With that in mind, let's get started!

### What Methodology to Employ

Four primary methodologies are commonly used in the supply-chain management process. Each of the methodologies has its own set of benefits to the user and each, correspondingly, has varying degrees of complexity. The methodology you choose is dependent on the specific needs of your organization, the commodities you purchase, and the resources available to implement and sustain your supply-chain management process.

When selecting the methodology (or methodologies) to be employed for your organization, consider first the scope and objectives that have been established for your supply-chain management process. To do so, several key questions need to be answered:

1. What is the intent of this process relative to reducing material costs?
2. What is the intent of this process relative to reducing supplier lead times?
3. What is the intent of this process relative to reducing the size of the supply base?
4. What is the intent of this process relative to improving incoming quality levels from the suppliers?
5. What is the intent of this process relative to reducing internal operating and inspection costs?
6. What resources have been allocated to this process? For what duration?
7. What budget has been approved for this process?
8. What is the expected return on investment from this process? When is it expected?
9. How many commodity categories currently comprise the supply base?
10. How many part numbers are included in each category?
11. How many suppliers by category comprise the supply base?
12. Based upon a standard Pareto analysis, what category or categories comprise 80% of the total volume of purchased materials by dollar expended?

Answers to these questions will guide you in establishing both the scope and objectives for your supply-chain management process. Remember, when establishing both the scope and objectives for your activities, be sure to align your expectations with the resources and budget allocated to you. Be realistic. Don't take on more than you can handle. It's better to start small and grow the process naturally than to start too big and see no results.

Another consideration is the restrictions placed on your organization by industry, customers, or regulatory bodies of federal, state, or local governments. For example, if your industry is governed by the FDA, certain requirements on your operation, and your supply chain, are mandated by law. If you ship products to the European Union, many of your operations, including how you source raw materials, may well be governed by the ISO category under which you are registered. Other restrictions could include those mandated by DoD, DOT, MIL Standards, CSA, UL, or QS 9000, to name just a few. It is therefore essential that those considerations and restrictions be considered in establishing both the scope and objectives for your process.

### The Four Methodologies

Time is critical. Resources are limited. Money is scarce. You know the drill, right? Your objective: conserve these limited commodities.

As we mentioned earlier, each of the four supply-chain management methodologies consumes differing amounts of those limited resources. So, based on your objectives, scope, and available resources, you must select the methodology that best addresses both your objectives and your constraints.

*Methodology One: The Survey Questionnaire*

The survey questionnaire is a methodology that is commonly used when resources and costs are an issue (and when are they not?). Useful for low dollar commodities and maintenance, repair, and office (MRO) items that have minimal impact on inventory levels, the survey questionnaire is an excellent screening tool to ensure that potential suppliers at least have the basic operational elements in place to meet your minimum quality and delivery requirements. The intent of the survey questionnaire is to:
- Provide information on supplier's capabilities, capacities, and process controls;
- Identify the products and services offered by the supplier;
- Highlight the supplier's cost structure; and
- Provide general logistic data on the supplier.

The scope of the survey questionnaire is limited to basic questions that relate to the supplier's general business operations and practices, along with its general financial condition. For each key question relating to process controls and capabilities, quality systems, and cost controls, documentation is requested to support the supplier's response. Each question focuses on specific targeted outputs, controls, and capabilities without being prescriptive in nature; each question, though, should be specific enough to yield usable information with which to fairly and objectively assess the supplier against its competition and your process objectives. And, it goes without saying, each question should be relevant, ethical, and legal.

A simple scoring system is applied to facilitate the processing of the questionnaires and the approval/disapproval of the supplier. Points are typically assigned to each question, depending on its relative level of importance, and a "floor" is established as a cutoff, below which the supplier is eliminated from further consideration.

Remember, the survey questionnaire is intended to provide hard, quantifiable evidence of the supplier's ability to comply with your supply-chain management process requirements. And even though it is limited in scope, it is still intended to provide a reasonable level of assurance that the supplier has the controls and capabilities you are seeking. So don't be shy about asking questions and seeking documented support for the things that are important to your organization. Do not accept evasive answers. Answers that relate to the supplier's *plans* are great, but they do not provide you with conforming products today, and there is no guarantee that they ever will. You are looking for what the supplier can do *today*.

Figure 8-1 shows an example of a survey questionnaire used by a major manufacturer of capital equipment to screen potential hardware and MRO suppliers. As you review it, assess its strengths and weaknesses.

*Methodology Two: Self-assessment Questionnaire*
A second, but more expansive, methodology to utilize when costs and resources are limited is the self-assessment questionnaire. This questionnaire is a useful tool when a long history exists with a given supplier and a significant amount of performance data has been collected to substantiate the supplier's capabilities and controls. As with the survey questionnaire, the intent of the self-assessment questionnaire is to save time and money by eliminating or minimizing the need for on-site audits of suppliers' processes and facilities. And as with the survey questionnaire, you must rely on the honesty and the integrity of the suppliers in answering the self-assessment questions, the accuracy and integrity of their data sources, and the capability and accuracy of their measurement systems. In short, you assume the burden of proof and all of the associated risks.

---

**ALLIANCE GROUP, INC.**
**2377 Anywhere Avenue, Suite 100**
**New York, New York 01245**

**(EXAMPLE ONLY)**

---

**SURVEY QUESTIONNAIRE**

This questionnaire is for the use of Alliance Group, Inc. (AGI) and its selected suppliers. The information gathered will be held in strict confidence by AGI, its employees, and assigns.

INSTRUCTIONS: All questions contained herein should be answered by the supplier. Any questions that do not pertain to the supplier's business operations should be identified as not applicable (NA). If the answer to a particular question is "none," please so state. Please be advised that we will be unable to process incomplete questionnaires. Attach additional sheets as necessary to address specific questions. Please complete one questionnaire for each facility that supplies products or services to AGI.

GENERAL INFORMATION

1. Company Name_____
2. Mailing Address_____
   _____
   _____
3. Telephone Number (_____) _____
4. Fax Number (_____) _____
5. Type of ownership_____
   (proprietorship, partnership, corporation, subsidiary, affiliate, division)
6. Length of time in business under this name_____
7. Length of time in business under another name_____
   Name_____
8. Parent company (if applicable)_____
9. Has company done business with AGI in the past?_____
   When?_____
10. Contact person_____ Telephone_____
11. Normal payment terms_____
12. Normal FOB point_____
13. Number of plant facilities servicing AGI_____
14. Size of facility(ies) servicing AGI (square feet/employees)_____
   _____
15. Employment per shift  (first)_____ (second)_____ (third)_____

---

*Figure 8-1. A supplier survey questionnaire kept current is a valuable tool to the supply-chain manager.*

AGI SURVEY QUESTIONNAIRE
PAGE TWO

LISTING OF KEY OPERATING PERSONNEL

President & CEO_____
Vice President–Operations _____
Vice President–Finance _____
Vice President–Marketing_____
Vice President–Engineering_____
Production Manager_____
Quality Manager_____
Customer Service Manager_____

16. Business category_____
    (large, medium, small, disadvantaged, US-owned, foreign owned, woman-
    owned, handicapped–not for profit)

Note: The term disadvantaged business designates a business that is at least
51% owned by one or more socially and/or economically disadvantaged individuals, or in the case of a publicly owned business, at least 51% of the stock is
owned by one or more socially and/or economically disadvantaged individuals,
and whose management and daily business operations are controlled by one
or more such individuals. For the purpose of this definition, socially and economically disadvantaged individuals include African Americans, Hispanic
Americans, Native Americans, Asiatic Americans, and other minorities, or any
individual found to be disadvantaged by the Small Business Administration
pursuant to Section (a) of the Small Business Act. Woman-owned business
designates a business that is at least 51% owned by a woman or women who
also control and operate the business. Control in this context means exercising
the power to make policy decisions affecting ongoing business operations. "Operate" designates day-to-day management.

17. How are products sold or distributed?_____
18. Products provided to AGI include_____
    _____
    _____

FINANCIAL INFORMATION

19. D&B rating_____
20. IRS identification number_____
21. Financial statements provided for year(s)_____
    (attached)

Figure 8-1. (Continued)

AGI SURVEY QUESTIONNAIRE
PAGE THREE

22. Annual sales revenues for last three years
    (YEAR)      (SALES REVENUES)

    _____   _____

    _____   _____

    _____   _____

23. Credit references we can contact
    (COMPANY)        (ADDRESS)                      (TELEPHONE)

    _____

    _____

    _____

    _____

24. Do you maintain a formal cost control system?_____
25. If so, has it been successful in reducing product or service costs to your
    customers during the past 12-24 months?_____

QUALITY MANAGEMENT

26. Are you a certified supplier for any other customer?_____
27. If so, who?_____
28. Are you ISO 9000 registered?_____
29. If so, under what category? (9001)___ (9002)____ (9003)____ (9004)___
30. What type of quality system do you employ?_____

    _____

31. Do you routinely provide certifications with your shipments of products?___
32. Do you have a written Quality Manual?_____
33. Do you use a formal supplier certification process in your procurement
    activities?_____
34. If a nonconformance is found by AGI or one of its customers, do you
    accept liability for:
    Product replacement?_____
    Labor costs to replace or reinstall?_____
    Lost time by AGI or its customers?_____
35. Do you routinely track customer service levels for:
    On-time delivery?_____ Quality?_____ Count Accuracy?_____

OPERATIONS MANAGEMENT

36. Do you maintain process controls within your operations that are statisti-
    cally based? _____

*Figure 8-1. (Continued)*

AGI SURVEY QUESTIONNAIRE
PAGE FOUR

37. Do you provide technical support to your customers relative to product design, process design, installation and field problems, product or service application issues, etc.? _____

38. If your primary business is distribution, do you have a formal distribution process control plan?_____

39. Is your company in compliance with all federal, state, and local regulatory requirements?_____

40. Do you maintain a formal inventory management system?_____

41. Are your labor relations stable? _____

42. Is your facility located near major transportation arteries? _____

43. Do you typically track the effectiveness and accuracy of your order entry process? _____

44. What is your typical customer order lead time for the products or services you provide?_____

ATTACHMENTS

Please attach documents you feel will assist AGI in evaluating your organization.

THE INFORMATION CONTAINED HEREIN IS COMPLETE AND ACCURATE TO THE BEST OF MY KNOWLEDGE AND BELIEF.

_____    _____

Signature of Authorized Representative                Title

_____    _____

Company Name                                         Date

PLEASE RETURN TO THE ATTENTION OF:

Ms. Jamie Jones, Supplier Certification Manager
Alliance Group, Inc.
2377 Anywhere Avenue, Suite 100
New York, New York 01245

or fax to: 212-345-6789

*Figure 8-1. (Continued)*

The self-assessment questionnaire is typically a scaled-down version of the full audit questionnaire, with the criteria and scoring mechanisms tailored to the particular commodity or commodities covered under the scope of this methodology. As before, responses from the supplier should be supported with quantifiable, documented evidence of conformance. Figure 8-2 is an example of a scoring system for a self-assessment questionnaire. Intentionally generic in nature, it can be tailored to your specific operation.

With this methodology, the burden of proof rests with the supplier, but the burden of assessment is your responsibility, and demands careful attention.

Shown in Figures 8-3 through 8-19 is a detailed self-assessment questionnaire developed by a recognized world-class apparel manufacturer. It is sufficiently comprehensive to provide the user with the evidence needed to assess the supplier's capabilities and controls.

*Methodologies Three and Four: Focused and Process Audits*
For the remainder of this chapter, we will focus on the last two methodologies, their development, and their deployment. These two methodologies represent the most effective approaches to yielding reliable results and, therefore, minimizing the level of risk in selecting the respective supplier(s). But because of their detail and scope, they are the most resource-intensive of the methodologies and command more time.

The process audit is a comprehensive assessment of all operational, support, and administrative activities that contribute to, influence, impact, or support those business processes associated with the design, production, and delivery of the supplier's products or services to the customer.

The focused audit, on the other hand, is a subset of the process audit, in which a review and analysis is done of specific elements of, or activities within, a given process such as quality management. It is not as broad-based, nor as resource-intensive as the process audit. It is usually used as a follow-up procedure to verify the implementation and effectiveness of corrective actions required as a result of a previous process audit. Other uses include the confirmation of specific self-assessment questionnaire responses to confirm their validity.

---

ABC MANUFACTURING, INC.

(EXAMPLE ONLY)

*Supplier Self-assessment Scoring System*

The following scoring scale has been provided for use by ABC's suppliers in evaluating their internal capabilities and in subsequently assigning a quantitative score which fairly and accurately represents their current level of operations. The purpose of this scale is to ensure that all suppliers are evaluated equally and consistently, with neither bias nor preconceived opinions relative to the process controls and capabilities demonstrated by each of ABC's suppliers.

| Audit Assessment | Points Awarded |
|---|---|
| All elements (100%) of this question are in compliance with ABC's requirements | 5 points |
| The majority of elements (90%) of this question are in compliance with ABC's requirements. Supplier is currently actively addressing the remaining elements. Documented programs are available to submit to ABC for review. | 4 points |
| Many elements (80%) of this question are in compliance with ABC's requirements. However, some improvement is required to bring the remaining elements into compliance. Supplier has initiated improvement efforts on several of the elements not currently in compliance. Said activities are documented and available for submission to ABC for review. | 3 points |
| Some elements (70%) of this question are in compliance with ABC's requirements. Much improvement is needed on those items currently not in compliance. Efforts underway to address those elements remain fragmented or poorly focused. Documentation is minimal. | 2 points |
| Only a few elements (60%) are in compliance with ABC's requirements. The supplier recognizes the need for improvement but has little documented evidence of actions taken or underway to address same. | 1 point |
| An insignificant number (<50%) of the elements of this question are in compliance. Little, if any activity is underway to address the identified improvements noted in this question. | 0 points |

*Figure 8-2. Scoring the self-assessment.*

| Self Assessment: Manufacturing Process Control | | | |
|---|---|---|---|
| | **Adequately Addressed** | **Not Adequately Addressed** | **Score (0-5)** |
| 1. Are all areas of the facility kept clean and free of nonessential inventory, tools, etc.? | | | |
| 2. Is there proper storage and control for hazardous materials used in the production process? | | | |
| 3. Is there a written procedure for electrostatic discharge protection when electrical components are used? | | | |
| 4. Do we utilize a "pull" versus "push" technique to drive production (i.e., kanban or final assembly order processing)? | | | |
| 5. Are manufacturing lots traceable throughout the production processes? | | | |
| 6. Is there a written procedure for statistical process control (SPC)?<br>  – Does it define reporting methods?<br>  – Does it outline the frequency and timing of samples?<br>  – Does it include the maintenance of statistically based control charts? | | | |
| 7. Are process controls established at all critical points within the process as defined by either the customers' or our process or design engineers? | | | |
| 8. Is process control data prepared and distributed often enough to provide early warning of developing process control problems? | | | |

*Figure 8-3.*

| Self Assessment: Manufacturing Process Control | | | |
|---|---|---|---|
| | Adequately Addressed | Not Adequately Addressed | Score (0-5) |
| 9. Does the process control data trigger corrective action when the process is not within control limits?<br>– Is the corrective action process clearly defined, including responsible parties?<br>– Can we demonstrate such actions? | | | |
| 10. Are process changes controlled, authorized, documented, dated, and signed? | | | |
| 11. Are process and product specifications readily accessible to operators? | | | |
| 12. Are operators knowledgeable and capable of interpreting customer specifications? | | | |
| 13. Are operators and maintenance personnel empowered to stop production operations?<br>– Under all out-of-control conditions?<br>– Is there a written procedure? | | | |
| 14. Are process and inspection records kept? For a specific period of time? | | | |
| 15. Are process and inspection records accessible to the operators? | | | |
| 16. Is there a written procedure for process audits?<br>– Does it specify methods for reporting findings and recommendations?<br>– Does it define methods of corrective action and responsibility? | | | |

*Figure 8-3. (Continued)*

| Self Assessment: Manufacturing Process Control | | | |
|---|---|---|---|
| | Adequately Addressed | Not Adequately Addressed | Score (0-5) |
| 17. Are SPC or inspection charts regularly reviewed by Manufacturing and Quality Management? | | | |
| 18. Do we have a documented rework procedure that is consistent with that of either the customer or applicable industry standards? | | | |
| 19. Is there a documented setup procedure? | | | |
| 20. Is there evidence of setup reduction activities to reduce setup time and costs? | | | |
| 21. Is production equipment calibrated on a regular basis by trained personnel? | | | |
| 22. Are calibration records and dates kept? | | | |
| 23. Are procedures in place to quarantine nonconforming materials? | | | |
| 24. Are procedures in place to confirm the acceptability of product for release to the customer? | | | |
| 25. Does an official engineering or spec change system exist that informs customers of changes to the products in advance? | | | |
| 26. Are all of our customers surveyed for approval of product changes before said changes are enacted? | | | |

*Figure 8-3. (Continued)*

| Self Assessment: Manufacturing Process Control | | | |
|---|---|---|---|
| | Adequately Addressed | Not Adequately Addressed | Score (0-5) |
| 27. Is there a written procedure for communication and distribution of specification and engineering changes? | | | |
| 28. Are obsolete engineering drawings and product specifications removed from all manufacturing operations immediately upon implementation of a new revision? | | | |
| 29. Are all documents and drawings used in the manufacturing process free of unofficial and handwritten changes? | | | |
| 30. Is there a properly documented and enforced preventive maintenance system? <br><br> – Predictive maintenance system? <br> – Are results available for review by senior management? | | | |
| 31. Are all tools and fixtures used in production fully qualified, maintained, and identified? | | | |
| 32. Do we employ a documented tool and fixture location system for all production tooling? | | | |
| 33. Is maximum tool life identified by the manufacturer and communicated to our operators? | | | |

*Figure 8-3. (Continued)*

| Self Assessment: Manufacturing Process Capability | | | |
|---|---|---|---|
| | Adequately Addressed | Not Adequately Addressed | Score (0-5) |
| 1. Do we conduct regular reviews of our process capability, with the intent to expand or enhance that capability? | | | |
| 2. Are the results of process capability studies forwarded to Design Engineering to be used in our product development activities? | | | |
| 3. Do we employ problem-solving techniques to identify, measure, and resolve internal and external process problems? | | | |
| 4. Are problem-solving techniques applied in a timely manner? | | | |
| 5. Is there a procedure that includes the use of problem-solving techniques to systematically reduce process variability with a goal of 100% first-pass yield? | | | |
| 6. Are statistical techniques used to continuously monitor process capability against product specifications? | | | |
| 7. Are comprehensive procedures available to our operators that define the actions to be taken to prevent our processes from moving out of control? | | | |
| 8. Do we employ design of experiments (DOE) or other preventive techniques to design out potential variability in our processes? | | | |
| 9. Do we require process capability studies from key suppliers prior to placing of orders with them? | | | |
| 10. Do we use the findings of these process capability studies in determining suppliers of critical components? | | | |

*Figure 8-4.*

| Self Assessment: Inventory Management Systems | | | |
|---|---|---|---|
| | **Adequately Addressed** | **Not Adequately Addressed** | **Score (0-5)** |
| 1. Do we have written procedures covering the storage, release, and handling of raw materials and inventory items? | | | |
| 2. Do these procedures ensure that only approved materials are released to and used in production? | | | |
| 3. Are in-process materials identified for tracking purposes? | | | |
| 4. Do we have an effective inventory control procedure that maximizes the inventory turns? | | | |
| 5. Does this procedure cover issues like shelf life and first-in-first-out methods to prevent deterioration of materials? | | | |
| 6. Do we use physical inventories or cycle counts to ensure inventory accuracy? <br> – What is current inventory accuracy? | | | |
| 7. Do we use a formal routing method to ensure that materials are produced to predetermined product or customer specifications? | | | |
| 8. Do we employ automated planning techniques to schedule and order raw materials and finished goods? | | | |
| 9. Do we label or code materials to ensure proper identification while they are in storage or in process? | | | |
| 10. Do our inventory control procedures effectively control surplus and obsolete inventories? <br> – What is our current level of surplus and obsolete inventory as a percent of total inventory? | | | |

*Figure 8-5.*

| Self Assessment: Quality Management | | | |
|---|---|---|---|
| | Adequately Addressed | Not Adequately Addressed | Score (0-5) |
| 1. Does our organizational structure support customer quality requirements as evidenced by a formal written quality policy?<br>– Measurable quality objectives?<br>– Performance metrics?<br>– Organizational structure with clearly defined lines of authority and responsibility for quality? | | | |
| 2. Does management promote operator control or in-process inspection by independent quality inspectors?<br>– Are operators adequately trained in the quality and technical aspects of their position? | | | |
| 3. Are inspection procedures monitored and enforced by management to ensure daily compliance? | | | |
| 4. Do test procedures and practices cover:<br>– How and when samples are taken?<br>– Equipment to be used for the test?<br>– Results obtained versus acceptance criteria?<br>– Methods of recording results?<br>– To whom results are to be reported?<br>– What action to take if results are not acceptable?<br>– How the material is released for further processing? | | | |
| 5. Has management organized regular quality review meetings? | | | |
| 6. Do management reviews encompass quality issues such as customer satisfaction, quality costs, internal audit reports, outgoing quality levels, etc.? | | | |

*Figure 8-6.*

| Self Assessment: Quality Management | | | |
|---|---|---|---|
| | Adequately Addressed | Not Adequately Addressed | Score (0-5) |
| 7. Do we have a long-term documented quality improvement plan? | | | |
| 8. Is achieving best-in-class status one of our quality improvement objectives? | | | |
| 9. Is the quality improvement plan broken down into individual departmental objectives with performance metrics to monitor progress? | | | |
| 10. Has management issued a written quality policy that is consistent with the quality improvement objectives of the company? | | | |
| 11. Is the quality policy communicated to all levels within the organization? | | | |
| 12. Is the quality policy understood by operating and support personnel? | | | |
| 13. Is the quality policy a "living document"; i.e., is the policy continually updated to reflect changes in customer require-ments, process improvements, etc.? | | | |
| 14. Does the supplier provide ongoing quality training for all levels and functions within the organization? | | | |
| 15. Is quality training documented in our employees' personnel records? | | | |
| 16. Is our quality system certified by an independent third party or customer? | | | |
| 17. Does our quality system include a follow-up process to monitor customer complaints and implement corrective actions? | | | |

*Figure 8-6. (Continued)*

| Self Assessment: Quality Management | | | |
|---|---|---|---|
| | Adequately Addressed | Not Adequately Addressed | Score (0-5) |
| 18. Does management regularly monitor such customer performance metrics as on-time delivery, incoming quality acceptance levels, and count accuracy? | | | |
| 19. Does management take corrective action when key customer performance metrics are not met? | | | |
| 20. Do we utilize statistical problem-solving methods to resolve performance problems?<br>– In all critical areas?<br>– Are results documented? | | | |
| 21. Do we use statistical process control (SPC) to ensure ongoing process control in both operating and support areas?<br>– Is SPC data available for review?<br>– Is SPC data used as the basis for corrective actions? | | | |
| 22. Do we promote a partnership relationship with both our customers and our suppliers? | | | |
| 23. Do we have a comprehensive quality manual? | | | |
| 24. Is the scope of our quality manual commensurate with the complexity of our products and those of our customers? | | | |
| 25. Are defective material reports (DMRs) used?<br>– Are DMRs used throughout our manufacturing processes, shipping, and receiving?<br>– Are DMRs used to initiate corrective action? | | | |

*Figure 8-6. (Continued)*

| Self Assessment: Quality Management | | | |
|---|---|---|---|
| | Adequately Addressed | Not Adequately Addressed | Score (0-5) |
| 26. Do we use quality stamps or certifications to indicate quality acceptance of products prior to shipment? | | | |
| 27. Are quality stamps or certifications controlled? | | | |

*Figure 8-6. (Continued)*

| Self Assessment: Management Commitment | | | |
|---|---|---|---|
| | Adequately Addressed | Not Adequately Addressed | Score (0-5) |
| 1. Is the company's organizational structure well defined and documented? | | | |
| 2. Does our management support and promote new ideas and concepts for continuous improvement? | | | |
| 3. Does our management provide employee training that is appropriate and relevant to individual job functions? | | | |
| 4. Are our training programs available to employees at all levels in the company? | | | |
| 5. Is management actively involved in the pursuit of quality and process improvements as evidenced by its participation in employee team activities? | | | |
| 6. Does management promote and support the concept of employee empowerment with a goal of self-directed work teams? | | | |
| 7. Has management identified and made provisions for the special controls, tools, skills, and processes required to ensure continuous product quality improvements? | | | |

*Figure 8-7.*

| Self Assessment: Management Commitment | | | |
|---|---|---|---|
| | Adequately Addressed | Not Adequately Addressed | Score (0-5) |
| 8. Has management established written goals and ongoing actions to reduce administrative cycle time for such processes as order entry, new product design, response to customer complaints, and requests for information? | | | |
| 9. Does management conduct a weekly review of company performance metrics such as on-time delivery, quality levels, count accuracy, order lead times, etc.? <br>− Does management initiate action when results do not meet objectives? <br>− Are the actions documented? | | | |
| 10. Are we doing business with any of our customer(s) as a partner or as part of a strategic alliance? | | | |
| 11. Is management receptive to innovation and improvement suggestions from our employees, suppliers, and customers? | | | |
| 12. Does management use the results of internal quality system audits to implement corrective actions? <br>− Are the actions documented? | | | |
| 13. Are customers notified of potential nonconformances or late deliveries in advance of the scheduled due date? | | | |
| 14. Does management personally visit customers and suppliers to solicit input about product and process improvements? | | | |
| 15. Does management give equal importance to administrative and product/service quality systems and practices? | | | |
| 16. Does management track cost-of-quality? | | | |

*Figure 8-7. (Continued)*

| Self Assessment: Management Commitment | | | |
|---|---|---|---|
| | **Adequately Addressed** | **Not Adequately Addressed** | **Score (0-5)** |
| 17. Does management have a current business plan?<br>– Companion product plan?<br>– Companion facility plan? | | | |
| 18. Does management promote employee involvement in professional organizations such as NAPM, ASQC, and APICS? | | | |
| 19. Has management developed a company mission statement that incorporates our customers' requirements? | | | |
| 20. Has management deployed its vision into all operating procedures? | | | |
| 21. Do operating procedures cover:<br>– How control is established in indirect and direct functions?<br>– How customer orders are to be controlled and processed?<br>– How product or service defects are to be addressed?<br>– How employees are to be trained?<br>– How products and processes are to be tested?<br>– How information is to be processed and controlled?<br>– Who is responsible for quality?<br>– Management's role in ensuring that quality requirements are met?<br>– Management's periodic auditing of the quality systems to ensure the continued suitability and effectiveness of the quality systems?<br>– What is expected from each functional area?<br>– Detailed instructions on how the policies and procedures are to be implemented?<br>– Handling of technical data, specs, and control parameters? | | | |

*Figure 8-7. (Continued)*

| Self Assessment: Business and Industry Knowledge | | | |
|---|---|---|---|
| | Adequately Addressed | Not Adequately Addressed | Score (0-5) |
| 1. Are our representatives and employees familiar with the customers' industries? | | | |
| 2. Does our strategic planning support our supplying products and services to all of the customers and industry segments we now service? | | | |
| 3. Are we active in industry trade groups, technical associations, or trade publications that support the industries we service? | | | |
| 4. Do we provide our customers with insight into market and industry trends and industry or competitive product developments? | | | |
| 5. Do we dedicate resources and investment to servicing our customers' markets? | | | |
| 6. Are we vulnerable to forces that would interfere with continued participation in the industries we now service? | | | |
| 7. Is our business dependent on just one or two customers? | | | |
| 8. Have we been awarded patents or royalties for products or technologies we have developed? | | | |
| 9. Are we working toward best-in-class status in the industries we service? | | | |

*Figure 8-8.*

| Self Assessment: Compliance of Purchased Materials | | | |
|---|---|---|---|
| | **Adequately Addressed** | **Not Adequately Addressed** | **Score (0-5)** |
| 1. Can we demonstrate that our raw materials, finished goods, and subcontracted services meet or exceed product specifications defined by our customers? | | | |
| 2. Do we confirm incoming quality of materials through incoming inspection or via supplier material certifications? | | | |
| 3. Do we have a supplier selection and certification program to qualify our suppliers? | | | |
| 4. Are supplier evaluations communicated to each supplier on a monthly or quarterly basis? | | | |
| 5. Do we require corrective action feedback on all defect items received from our supply base? | | | |
| 6. Are written specifications required for all incoming materials, and are they complete? | | | |
| 7. When we utilize subcontractors for in-process operations, are there controls in place to ensure our quality requirements are met? | | | |
| 8. Are complete records maintained on our approved suppliers? | | | |
| 9. Have quality, on-time delivery, and count accuracy metrics been established, and are they being actively monitored for each supplier? | | | |
| 10. Do we use a supplier rating system, and are suppliers informed of their performance? | | | |

*Figure 8-9.*

| Self Assessment: Compliance of Purchased Materials | | | |
|---|---|---|---|
| | **Adequately Addressed** | **Not Adequately Addressed** | **Score (0-5)** |
| 11. Are our suppliers audited by our internal employees or independent auditors on a regular basis? | | | |
| 12. Do we maintain data on each supplier's process capabilities? | | | |
| 13. Do our suppliers provide samples or test data with specifications prior to our placing a purchase order? | | | |
| 14. Is there a written procedure covering the receipt of materials, and is the procedure audited regularly? | | | |
| 15. Is incoming material properly segregated from previously accepted materials until the approval process is completed? | | | |
| 16. Is the incoming material properly protected from the environment upon receipt? | | | |
| 17. Do we have a procedure that provides identification and traceability of incoming lots? | | | |
| 18. Are raw-material testing procedures properly documented? | | | |
| 19. Are raw-material test results monitored by management-level employees? | | | |
| 20. Do we have electronic data interchange (EDI) capability? | | | |

*Figure 8-9. (Continued)*

| Self Assessment: Cost Controls | | | |
|---|---|---|---|
| | **Adequately Addressed** | **Not Adequately Addressed** | **Score (0-5)** |
| 1. Is the percentage of goods and services sold to any one customer in excess of half the total goods and services sold to all customers? | | | |
| 2. Can we demonstrate that we are actively involved in cost containment and reduction programs in the areas of: <br>– Waste reduction? <br>– Supplier cost controls? <br>– Productivity improvements? <br>– Efficiency improvements? <br>– Technological advancements? <br>– Product development? <br>– Administrative cycle time? | | | |
| 3. Have we established cost standards against which operational metrics can be applied? | | | |
| 4. Are operational metrics tracked daily and variances to standards reviewed for initiation of corrective actions? | | | |
| 5. Are cost standards compatible with all quality, operational, and product metrics? | | | |
| 6. Do we appropriately apply indirect and overhead costs to the products and services provided to our customers? | | | |
| 7. Do our accounting functions interact frequently and supportively with our operational functions? | | | |
| 8. Do we track our cost of quality? | | | |
| 9. Are corrective action plans implemented from our cost-of-quality measurements? | | | |
| 10. Do we effectively control overtime and other lead time-associated costs? | | | |

*Figure 8-10.*

| Self Assessment: Cost Controls | | | |
|---|---|---|---|
| | Adequately Addressed | Not Adequately Addressed | Score (0-5) |
| 11. Do we effectively control inventory-related costs (raw materials, work-in-process [WIP], finished goods)? | | | |
| 12. Are our labor costs controlled effectively? | | | |
| 13. Are we willing to share cost information with our customers for the products and services we provide? | | | |
| 14. Do we include an internal interest rate or cost-of-capital figure in our cost accounting system? | | | |
| 15. Can we demonstrate that our cost controls have been successful in reducing product or service costs during the prior 12-24 months? | | | |
| 16. Are we willing to work with our customers in a partnership arrangement that focuses on mutual benefit from the reduction of product and service costs (i.e., 50-50 split of all cost reduction results)? | | | |

*Figure 8-10. (Continued)*

| Self Assessment: Order-entry Process | | | |
|---|---|---|---|
| | Adequately Addressed | Not Adequately Addressed | Score (0-5) |
| 1. Are written procedures in place covering the order-entry process? | | | |

*Figure 8-11.*

| Self Assessment: Order-entry Process | | | |
|---|---|---|---|
| | Adequately Addressed | Not Adequately Addressed | Score (0-5) |
| 2. Do the procedures include key customer-specific performance metrics? | | | |
| 3. Does management monitor the metrics to ensure accurate and timely processing of customer orders? | | | |
| 4. Do we know the current order-entry accuracy rate? | | | |
| 5. Do we know the target accuracy rate? | | | |
| 6. Do we know the current order-entry cycle time? | | | |
| 7. Do we know the target order-entry cycle time? | | | |
| 8. Does Order-entry monitor:<br><br>– Customer satisfaction levels?<br>– On-time delivery levels?<br>– Product quality levels?<br>– Customer complaint resolutions?<br>– Pricing errors?<br>– Invoicing errors? | | | |
| 9. Are the results of these customer-specific performance metrics used by management to implement continuous improvement processes? | | | |
| 10. Can we demonstrate how these actions have led to improved customer service levels? | | | |

Figure 8-11. (Continued)

| Self Assessment: Customer Service | | | |
|---|---|---|---|
| | Adequately Addressed | Not Adequately Addressed | Score (0-5) |
| 1. Can we demonstrate that our customer service function is clearly defined with a distinct organizational structure? | | | |
| 2. Is a measurement system in place to effectively and consistently evaluate our customers' satisfaction levels? | | | |
| 3. Are the metrics we use to measure customer satisfaction accurate and complete? | | | |
| 4. Is our senior management actively involved in customer service? | | | |
| 5. Are both positive and negative trends in customer satisfaction reported directly to our senior management? | | | |
| 6. Do we maintain a high level of customer service throughout our entire customer base? | | | |
| 7. Do we compare favorably with our competition relative to customer satisfaction indices? | | | |
| 8. What has the trend been in our customer satisfaction over the last 3-5 years? | | | |
| 9. Have customer satisfaction goals been established for our organization? | | | |
| 10. Have we benchmarked our customer satisfaction goals against best-in-class organizations within our industry(ies)? | | | |
| 11. When customer satisfaction levels fall short of targeted or expected goals, does senior management direct corrective action efforts? | | | |

*Figure 8-12.*

| Self Assessment: Customer Service | | | |
|---|---|---|---|
| | Adequately Addressed | Not Adequately Addressed | Score (0-5) |
| 12. In the last 6 months, has our senior executive personally spoken with employees or executives of our primary customers (i.e., those that collectively comprise 80% of our annual sales revenues)? | | | |
| 13. Does our customer service consistently provide prompt (within 24 hours) resolution of customer complaints and/or requests for information? | | | |
| 14. Do we have and follow a documented procedure for handling customer complaints? | | | |
| 15. Is the procedure comprehensive? | | | |
| 16. Is the procedure followed by all customer service staff at all times? | | | |
| 17. Are customer service staff empowered to resolve customer complaints on the spot, without seeking management approval? | | | |
| 18. Is there a procedure to advise customers of potential delivery, quality, or service problems in advance? | | | |
| 19. Is there a procedure to advise customers of manufacturing or process changes in advance? | | | |
| 20. Are we flexible in both delivery schedules and order quantities? | | | |
| 21. Do we provide field support when needed to resolve operational or installation problems? | | | |

*Figure 8-12. (Continued)*

| Self Assessment: Customer Service | | | |
|---|---|---|---|
| | **Adequately Addressed** | **Not Adequately Addressed** | **Score (0-5)** |
| 22. Do we offer our customers EDI capability? | | | |
| 23. Are we actively pursuing lead-time reduction activities? | | | |
| 24. Can we demonstrate that our lead-time reduction efforts have yielded positive results for our customers? | | | |
| 25. Can we demonstrate that our management and employees are dedicated to ethical business practices? | | | |

*Figure 8-12. (Continued)*

| Self Assessment: Labor Relations | | | |
|---|---|---|---|
| | **Adequately Addressed** | **Not Adequately Addressed** | **Score (0-5)** |
| 1. Do we promote employee involvement activities such as functional or cross-functional work teams? | | | |
| 2. Do we provide training and educational opportunities for all employees at every level within the organization? | | | |
| 3. Do we maintain training records on each employee, illustrating the courses and educational opportunities afforded each employee? | | | |
| 4. Are we an equal opportunity employer? | | | |
| 5. Have all labor disputes been resolved either with or without outside intervention or mediation? | | | |

*Figure 8-13.*

| Self Assessment: Labor Relations | | | |
|---|---|---|---|
| | Adequately Addressed | Not Adequately Addressed | Score (0-5) |
| 6. Does supplier support an active employee suggestion program? | | | |
| 7. Are employees involved actively in company decision-making? | | | |
| 8. Does our senior management keep employees advised of financial and market conditions of the company? Frequently? | | | |
| 9. Are employees involved in the establishment of the company's operational, market, product, and financial objectives? | | | |
| 10. Can we demonstrate that we can equal or better industry averages for:<br><br>– Employee retention rate?<br>– Employee absenteeism?<br>– Employee productivity?<br>– Employee advancement? | | | |
| 11. Do we have any pending or active litigation relating to labor disputes, sexual harassment, discrimination, or unfair labor practices? | | | |
| 12. Is the average length of service for our employees comparable with other companies in our industry? | | | |
| 13. Do we have a documented, active cross-training program to broaden employee skill sets? | | | |

*Figure 8-13. (Continued)*

| Self Assessment: Compliance with Governmental Regulations | | | |
|---|---|---|---|
| | Adequately Addressed | Not Adequately Addressed | Score (0-5) |
| 1. Can we demonstrate that we are in full compliance with all federal, state, and local regulatory requirements applicable to our particular industry segment (EEOC, OSHA, EPA, etc.)? | | | |
| 2. Do we have documented procedures in place that ensure environmentally-responsible and community-supportive (proactive) operations? | | | |
| 3. Can we demonstrate that any products provided to our customers are in compliance with applicable regulations, and that any products containing materials or chemicals covered under restrictions are labeled appropriately, pursuant to applicable regulatory requirements? | | | |
| 4. Have we been cited within the last 3 years or is litigation pending for noncompliance with federal, state, or local environmental or employment regulations? | | | |
| 5. If no. 4 is affirmative, has corrective action been initiated to effectively address the noncompliance? | | | |
| 6. Do we maintain documented safety and health awareness programs as part of our normal operations? | | | |
| 7. Is our safety and health awareness program audited annually to determine its effectiveness? | | | |
| 8. Are all personnel trained relative to safety, health, and other governmentally-sanctioned regulatory requirements? | | | |

*Figure 8-14.*

| Self Assessment: Compliance with Governmental Relations | | | |
|---|---|---|---|
| | **Adequately Addressed** | **Not Adequately Addressed** | **Score (0-5)** |
| 9.  Do we have a documented substance abuse program/policy? | | | |
| 10. Do we have emergency procedures to respond to crisis situations? | | | |
| 11. Are Material Safety Data Sheets (MSDS) and other safety informational documents provided to employees? | | | |
| 12. Are there documented procedures in place covering the design, development, control, and distribution of environmentally-sensitive and hazardous materials? | | | |
| 13. Do we have an active recycling program in conjunction with both our suppliers and customers? | | | |

*Figure 8-14. (Continued)*

| Self Assessment: Facilities Management | | | |
|---|---|---|---|
| | **Adequately Addressed** | **Not Adequately Addressed** | **Score (0-5)** |
| 1.  Do housekeeping procedures exist in every department? | | | |
| 2.  Are housekeeping audits scheduled and conducted on a regular basis? | | | |
| 3.  Is there a preventive and predictive maintenance program in place for all operating areas and facilities? | | | |
| 4.  Are preventive maintenance schedules adhered to by workers in all areas and at all levels of management/supervision? | | | |

*Figure 8-15.*

| Self Assessment: Facilities Management | | | |
|---|---|---|---|
| | Adequately Addressed | Not Adequately Addressed | Score (0-5) |
| 5. Are there written procedures for reporting deviations from the preventive maintenance standards? | | | |
| 6. Are records maintained for each piece of equipment, and are records accessible to operators and maintenance personnel? | | | |
| 7. Is the responsibility for the preventive maintenance program clearly defined? | | | |
| 8. Are all material storage areas clean and clear of excess materials, tools, fixtures, and packaging dunnage? | | | |
| 9. Are all manufacturing areas clean and clear of excess materials, fixtures, tools, and packaging dunnage? | | | |
| 10. Do housekeeping, safety, and hygiene practices meet industry standards? | | | |
| 11. Is there a formal safety review program for production, installation, and storage areas? | | | |
| 12. Is timely corrective action taken, based on records of work injuries and lost-time accidents? | | | |
| 13. Are there written safety instructions (safety manuals, warnings, etc.)? | | | |
| 14. Are there written safety instructions for special processes such as air conditioning, hazardous waste treatment, etc.? | | | |
| 15. Is there a written procedure for electrostatic discharge protection where applicable? | | | |

*Figure 8-15. (Continued)*

| Self Assessment: Document Control Processes | | | |
|---|---|---|---|
| | **Adequately Addressed** | **Not Adequately Addressed** | **Score (0-5)** |
| 1. Do we maintain a document control procedure that ensures that our customers' specifications, bills of material, engineering change notices, and drawing revision levels are current and available to all operating personnel? | | | |
| 2. Are all associated employees trained on the document control procedure? | | | |
| 3. Are these procedures audited on a periodic basis to ensure compliance? | | | |
| 4. Are the results of the audits available for review by senior management? | | | |
| 5. Does Order Entry check documentation prior to the processing of each order from our customers? | | | |
| 6. Is design and process documentation complete? Current? | | | |
| 7. Is there a written procedure that covers the retention period for key documents? | | | |
| 8. Is there a storage and retrieval procedure for all key documents? | | | |
| 9. Is there an off-site storage location for key documents and backup copies? | | | |
| 10. Are all document storage locations secure with controlled access? | | | |
| 11. Are we seeking ISO 9000 registration? | | | |

*Figure 8-16.*

| Self Assessment: Distribution Process Control | | | |
|---|---|---|---|
| | Adequately Addressed | Not Adequately Addressed | Score (0-5) |
| 1. Do we maintain packaging specifications? | | | |
| 2. Is responsibility for packaging specifications clearly defined? | | | |
| 3. Are written procedures employed in our packaging and shipping departments? | | | |
| 4. Do our procedures allow packages to be specified by our customers? | | | |
| 5. Are there written procedures for marking containers for shipping in accordance with customer specifications? | | | |
| 6. Are proper identification labels used for each package or container of a shipment? | | | |
| 7. Are customer routing and traffic instructions visible on the product and within the shipping and packaging departments? | | | |
| 8. Do we employ and offer bar coding to our customers? | | | |
| 9. Is the responsibility for our company logistics function clearly defined? | | | |
| 10. Do we negotiate agreements with our freight carriers to ensure on-time delivery at the most advantageous rate to our customers? | | | |
| 11. Is a certificate of compliance to customer specifications and test requirements enclosed with all shipments to customers? | | | |

*Figure 8-17.*

| Self Assessment: Distribution Process Control | | | |
|---|---|---|---|
| | Adequately Addressed | Not Adequately Addressed | Score (0-5) |
| 12. Are customer part numbers, work order numbers, purchase order (PO) numbers, and material codes marked on outgoing containers for ease in identification? | | | |
| 13. Are approved lots kept intact and segregated throughout the shipping and packaging processes? | | | |
| 14. Is nonconforming material properly identified and segregated from qualified material? | | | |
| 15. Is storage and release of materials for shipping restricted to only authorized personnel? | | | |
| 16. Is there a shelf-life program for distributed products? | | | |
| 17. Are written agreements in place with transport companies and forwarding agents that provide for the return of defective or faulty materials? Is there a process for tracking material not received? | | | |
| 18. Do we have a proceduralized returned-goods policy? | | | |
| 19. Do we support the customers' requested FOB point? | | | |
| 20. Do we offer ship-direct services, and are they proceduralized? | | | |
| 21. Is there a procedure to track shipments from point of origin to destination? | | | |
| 22. Is material that is ordered for same-day shipment (UPS, Federal Express, etc.) documented as to actual date shipped? | | | |

Figure 8-17. (Continued)

| Self Assessment: Product Warranty and Reliability | | | |
|---|---|---|---|
| | Adequately Addressed | Not Adequately Addressed | Score (0-5) |
| 1. Can we demonstrate that product-reliability and life-cycle costs have been confirmed through Failure Mode and Effects Analysis (FMEA), Design of Experiments (DOE), or other statistically-based methodologies? | | | |
| 2. Do we routinely utilize "design for producibility, assembly, and maintainability" methodologies? | | | |
| 3. Are these methodologies included in our design rules and documented as normal operating practice? | | | |
| 4. Do we routinely track product life cycles and life-cycle costs? | | | |
| 5. Do our products conform to advertised specifications, performance levels, and life-cycle costs under our customers' applications and environment? – Is there data to support our position? | | | |
| 6. Do our product warranties cover all applications, conditions, and environ-ments experienced by our customers? | | | |
| 7. Are our warranty periods compatible with those of our customers' products? | | | |
| 8. Do we have a formal procedure for handling warranty claims? | | | |
| 9. Can we provide documented evidence of adherence to our warranty policy under normal conditions? | | | |
| 10. Are our warranty claims processing procedures available for review by our customers? | | | |

*Figure 8-18.*

| Self Assessment: Technical Support | | | |
|---|---|---|---|
| | **Adequately Addressed** | **Not Adequately Addressed** | **Score (0-5)** |
| 1. Is the scope of our R&D and product development activities sufficient for the industries we service? | | | |
| 2. Are our R&D and new product development efforts focused on meeting the stated needs of our customers? | | | |
| 3. Are procedures in place that ensure that our customers' specifications are used in the new-product development process? | | | |
| 4. Do we employ concurrent/simultaneous engineering methodologies in the development of new products? | | | |
| 5. Is our new-product development process clearly defined and proceduralized? | | | |
| 6. Do our engineers incorporate internal manufacturing process capabilities into the design rules/criteria for new products? | | | |
| 7. Are design changes effectively controlled throughout the organization? <br> – Is the process proceduralized? <br> – Are engineering changes kept to a minimum (less than three per quarter)? | | | |
| 8. Are engineering and R&D personnel active in professional societies and continuing education? | | | |
| 9. Are computer-based design tools used in the design of new products and processes? | | | |
| 10. Is computer simulation and modeling used to maximize the reliability and life cycle of new products? | | | |

*Figure 8-19.*

| Self Assessment: Technical Support | | | |
|---|---|---|---|
| | Adequately Addressed | Not Adequately Addressed | Score (0-5) |
| 11. Do we use FMEA and DOE methods to maximize product reliability while reducing manufacturing and life-cycle costs? | | | |
| 12. Do we employ a policy for patents to encourage innovation? | | | |
| 13. Does a long-term technology plan exist to guide continued development of leading-edge technologies? | | | |
| 14. Do we have an effective system for managing and storing customer-supplier documentation? | | | |
| 15. Do we have a mechanism for notifying customers of documentation problems? | | | |
| 16. Do new product/process development procedures define metrics to be employed to ensure compliance with design rules, cycle times, performance requirements, reliability requirements, etc.? | | | |
| 17. Is there a procedure to control critical-to-function characteristics for new products before production, including SPC characteristics and frequencies? | | | |
| 18. Are key suppliers included as part of the design team? | | | |
| 19. Are key customers included as part of the design team? | | | |

Figure 8-19. (Continued)

Remember, the objective of each of these methodologies is to confirm the supplier's:
- Process *capability*,
- Process *controls*, and
- *Commitment* to continuous conformance to your specific requirements.

Let data be your driver. Adopt a "show-me" attitude.

### Getting Started

Like any other strategic initiative, supply-chain management requires a high level of planning and preparation. As we discussed earlier, you must first establish the scope and objectives for your process. Next, you must develop the performance metrics you will use to measure the conformance of your suppliers to the requirements you have established for your process. And finally, you must develop a checklist of the required actions, tools, resources, timing, and support you will need to deploy your strategies.

We have discussed the setting of the scope earlier in this chapter. A few additional comments, however, are in order.

Be sure your scope is definitive. In other words, specify what is included and what is *not* included in your scope. Be sure you receive the buy-in from all functional managers required to support your supply-chain initiative, as well as from senior management. And finally, be sure that the resources you will need have been committed to you and that they will be deployed when and where you need them. To support your supply-chain management initiative, develop a formal project plan that includes both a detailed schedule and budget for your initiative, and then get approval from management for those resource requirements.

During this phase of the process, it is advisable to develop an Objectives Matrix (as illustrated in Figure 8-20) to assist you in quantifying the objectives for your initiative. It forces you to be specific and, by so doing, adds a degree of reality to the setting of those objectives. Remember, the objectives you set must be consistent with the scope of your initiative and the strategic direction of your organization, and be realistic given the resources and timing allotted for your supply-chain management process.

Before moving forward, it is advisable that you fill in your checklist to ensure that you haven't missed anything (Figure 8-21).

| OBJECTIVES MATRIX | | | | | |
|---|---|---|---|---|---|
| Objective | By When? | By How Much? | Expected ROI | Metrics | Data Sources |
| Improve quality level | July, 1995 | to 95.0% | 32% | IQL | IRR report, |
| | Dec, 1995 | to 97.5% | +26% | WIP | Variance |
| | July, 1996 | to 98.5% | +24% | Warranty | reports |
| Improve deliveries | July, 1995 | to 97.5% | 37% | -5/+5 day | PO receipts, |
| | Dec, 1995 | to 99.5% | +41% | -2/+0 day | Receiving |
| Reduce lead time | Dec, 1995 | to an avg of 3 wks | 54% | Days from OE to ship | PO status reports |
| Reduce supply base | Dec, 1995 | 350 to 325 | N/A | Number | Active, |
| | July, 1996 | 325 to 275 | | Total & by commodity | A/P report |

*Figure 8-20. An Objectives Matrix provides a discipline for quantifying objectives.*

| PREPARATION CHECKLIST | | | | |
|---|---|---|---|---|
| REQUIRED | Actions | Tools | Resources | Timing |
| What | | | | |
| When | | | | |
| Confirmed | | | | |
| What | | | | |
| When | | | | |
| Confirmed | | | | |
| What | | | | |
| When | | | | |
| Confirmed | | | | |

*Figure 8-21. A Preparation Checklist will reveal whether you are ready to proceed with your supply-chain management initiative.*

### Defining the Baseline Requirements

The next step in the supply-chain management process is to develop a listing of requirements that any supplier must provide or possess in order to do business with your organization. This becomes the baseline for all future selection and qualification processes. For example, your quality system may be based on statistical process control, design of experiments, FMEA, or other statistically-based quality techniques. If that be the case, it may be a requirement that all incoming materials conform to those same rigid statistical standards. As such, one of your requirements for any supplier of critical materials could be that they, too, must use similar statistically-based quality techniques.

As in the alternatives analyses discussed in earlier chapters, these "must" requirements become the filters for our decision-making process. Any supplier that does not meet these essential requirements cannot be considered for a long-term relationship with your firm, and should therefore not be considered further.

In many organizations today, the ISO 9000 and QS 9000 guidelines are used as the basis for the supply-chain management process. This ensures that consistency is maintained with these two international standards, while providing a framework that is recognized throughout the supply chain of most industries. The examples provided in this chapter will thus be tailored around ISO 9000 and QS 9000.

Again, your needs and requirements may vary from those presented here. It is therefore essential that the following be used only as a rough set of guidelines for the development of a process that fits your own unique environment and organizational constraints. We begin by looking at several of the "must" requirements used by many of the world-class supply-chain management programs.

### Manufacturing Process Capabilities and Capacities

Every supplier must be able to clearly demonstrate that their internal processes are statistically capable of meeting or exceeding the specifications and requirements of your organization's products and services. All of the supplier's critical process capabilities should be documented through statistically-based process capability studies that take into consideration the manufacturer's recommended tolerances, actual operating conditions, and the physical condition of the manufacturing equipment (as supported by active preventive and predictive maintenance procedures).

Practical capacity must be documented and supported by effective capacity management systems to monitor peak loading on critical machining or bottleneck centers. The supplier's work force must be capable of moving between machining or process centers to address throughput issues without adding operating overhead, delaying shipments, or creating backlogs. Critical measures of manufacturing cycle time, throughput, and operating expense must be used as the primary drivers for the manufacturing process.

## Manufacturing Process Control

All existing and potential suppliers must be able to demonstrate that they have in place documented and active in-process manufacturing control procedures to verify that the critical quality characteristics identified on your organization's prints and product specifications are maintained through statistical methodologies. The data necessary to support the supplier's manufacturing control systems and procedures should be obtained and assessed on a regular basis, with all corresponding corrective actions documented to ensure that all process and product nonconformances are resolved as evidenced by tangible, measurable results. As with all quality-related issues, the supplier's manufacturing process controls should focus on prevention versus appraisal techniques to keep operating expenses and manufacturing cycle times at their lowest.

In addition, manufacturing process control procedures should be in place to test and confirm that production and assembly processes are conducted under controlled conditions, including the use of approved equipment, inspection techniques, and associated gage calibration methodologies. In addition, there must be evidence that the supplier supports their manufacturing process controls through formal instruction and training programs for all administrative, direct, and indirect employees.

## Quality Systems

Each of your key suppliers should be capable of demonstrating that they employ formal quality systems that ensure continuous process control throughout all administrative, direct, and support functions. Each element of the supplier's quality system should be formally documented and encompass all processes that contribute to the receipt, production, handling, storage, and distribution of the products and services provided to your organization, or distributed directly to your organization's customers. The supplier's quality systems should be based on recognized statistical methodologies that comply with QS 9000, ISO, ANSI, ASTM, UL, CSA, FDA, or other applicable standards. The process controls and capabilities governing the supplier's quality systems must guarantee that all functional product specifications and requirements are deployed throughout every level within your supplier's operations, as well as throughout all functional activities.

For example, your supplier's quality system should clearly state who is responsible for the control, containment, and release of all nonconforming materials, and under what conditions those nonconforming materials can be released for further processing. The documentation used to control all nonconforming materials should clearly describe what criteria are to be used to define how nonconforming materials are to be identified, how they are to be segregated, and what evaluation techniques or methods are to be used. Your supplier's quality system should guarantee that any in-process materials identified at any stage of the process as not conforming to the specifications of your organization, must be either scrapped or reworked under guidelines defined and approved by your firm. In addition, your supplier's quality system must isolate the root cause of the nonconformance in administrative, operating, or support functions, as well as through the supplier's own supply chain. It must then define the corrective actions required to guarantee that the same nonconformance will not occur again in the future.

As a second example, your supplier's quality system must ensure that all confirmations and measurements are taken with quality tooling and instrumentation of known accuracy and calibration. Your supplier's quality system should also clearly define the frequency of calibration for each quality tool, as well as the selection, control, and maintenance processes and methodologies for all quality assurance and test equipment. It is also a good idea for your supplier's quality system (and your own) to include the requirement for periodic internal audits to confirm that all applicable quality activities comply with their customers' requirements, and to assess the effectiveness of the supplier's quality system in meeting or exceeding those same customer expectations. The results of the supplier's audits should be documented and, for maximum effectiveness, reviewed or audited by the supplier's senior management to ensure that any required corrective actions have been effective in eliminating any defects.

## Quality Management

Your suppliers should be able to demonstrate that their organizations are structured to ensure a focus on providing quality products and customer service. Evidence of such a customer-focused organization can be obtained through a formal written quality policy, measurable quality objectives and performance metrics, and an organizational structure that clearly defines the lines of authority and responsibility for quality. Rather than the typical inspection-focused quality systems, the supplier's senior and operating-level management must promote both operator and process control to ensure continuous process and product conformance. In addition, the supplier's senior and operating-level management must provide all operators and quality technicians with the necessary technical training to support the particular requirements of their job or function. Further, the supplier's quality management should regularly verify that comprehensive procedures for statistical process control, in-process inspections or confirmations, raw material receiving certifications or confirmations, and final quality acceptance methodologies are in place *and understood* by all operators and quality personnel. The supplier's quality management must also verify that all required tests, inspections, critical characteristics, and statistical control points are clearly defined, including the responsible party(ies) and the frequency of confirmation.

For example, the supplier's quality management procedures should cover:

- How, when, and where to take samples;
- What equipment and gaging is to be used for testing and statistical sampling;
- How to compare test results against acceptable criteria;
- What methods to employ in recording data and test results;
- To whom results are to be reported and how often;
- What actions to take in the event results fall outside acceptable control limits; and
- How materials (conforming and nonconforming) are to be released for additional processing.

*Compliance of Raw Materials*
The supplier must be able to provide evidence of the procedures and practices used by their purchasing operation to acquire raw materials, finished goods, and/or subcontracted services that specifically meet or exceed the specifications required by your organization for your products and services. That type of compliance must be confirmed through documented supplier assessments, audits, and performance monitoring techniques quantified through formal measurement and assessment methodologies. The supplier's corrective action procedures covering product or process nonconformances should be clearly documented and understood by all operating and quality personnel, and the results of those corrective actions documented. In general, all purchasing procedures describing how raw material or component purchase orders are placed, controlled, and authorized should be clearly documented and followed in daily practice.

*Inventory Control Processes*
Every existing or potential supplier should be capable of demonstrating that their inventory control procedures accurately describe how all raw, in-process, and finished goods inventories are stored, handled, and packaged. Inventory storage procedures must address shelf-life issues (where applicable), contamination, electrostatic control, and other material quality issues that could impact the material's suitability for production or subsequent processes. The supplier's procedures should define the type of package materials to be used for storage, handling, and shipment; any labelling requirements dictated by your organization; applicable bar coding formats and requirements; as well as any environmental impact requirements.

All inventory order planning, processing, and release procedures should be clearly defined by the supplier's procedures, including all inventory planning criteria that are in place to prevent inventory obsolescence and surpluses. Dynamic measures of inventory performance, like turns or percent of sales, should be used by the supplier to monitor its return on inventory investment and corresponding cash flow impacts.

*Technical and Product Support*
Existing or potential suppliers should be able to demonstrate the ability to provide timely no-cost or low-cost technical support to

enhance your organization's current and planned product lines, aftermarket and repair services, and any anticipated growth in market share, sales, or technological innovation. That support should be provided by technically qualified sales engineers/representatives, applications engineers, supplier quality engineers, manufacturing engineers, and/or design engineers with a focus on assisting your organization with:

- Problem resolution,
- Product planning,
- Research and development support for your design engineers in new product design and development through a concurrent engineering methodology,
- Support in developing technical instruction manuals and installation guides for the products and services provided to your organization, and
- Any product-related customer training on an "as needed" basis.

## Cost Controls

Suppliers should be able to demonstrate an active involvement in cost-containment programs that have yielded consistently positive results in maintaining or reducing the product or service costs to their customers. Demonstrated results should be defined as measurable cost reductions evidenced by the suppliers without an offset in either quality or delivery performance. Suppliers should also be willing to cooperate with your organization to share equally the results of any such cost reductions, as well as to work with your organization to develop or enhance cost-reduction or containment programs for the benefit of both parties.

## Commitment of Management

The commitment of the supplier's management to serving their customers should be evident by the incorporation and deployment of customers' expectations and requirements into the organization's mission statement, as well as all company operating procedures. Company operating procedures should document how all company functions and processes are to be managed and operated to meet those customer requirements and expectations. Such issues and areas that should be documented include:

- How control is established and maintained throughout all administrative, direct, and indirect functions;

- How customer orders are to be processed and controlled, and what performance metrics will be used to ensure continued compliance to expected performance levels;
- How product or service deficiencies are to be addressed, once identified;
- How employees are to be trained, on what topics, and with what frequency;
- How primary and critical products and processes are to be tested and validated, and what methods will be used to do so;
- How vital quality- and manufacturing-related information will be processed and controlled, and who will have accountability for it;
- A clear definition of who is responsible for quality, as well as the technical competency requirements for it;
- The role senior management plays in ensuring that all customer quality requirements are met and maintained;
- Procedures covering management's periodic auditing of the supplier's quality system to ensure the continued suitability and effectiveness of the system;
- A comprehensive procedure or process map that delineates specifically what is expected from each functional discipline;
- Detailed instructions defining how all company policies and procedures are implemented, monitored, and maintained; and
- Accurate technical data, control parameters, and specifications of the products and services provided to customers.

The supplier's senior management must be able to demonstrate that the level of expertise required for each critical function has been accurately defined, and that the people performing those functions meet or exceed those requirements. Training records should be available to demonstrate that the specified employee capabilities and skill levels exist. In addition, records should be kept indicating that the training and education provided by the supplier or obtained for the employees relative to those critical skill sets has been effective in reducing quality defects and manufacturing cycle time.

*Financial Stability*
Every supplier should be capable of demonstrating that its company is financially sound and effectively utilizing all of its assets to

generate cash flows from internal operations that are sufficient to sustain continuing business operations. Additionally, the supplier should guarantee that if those financial conditions change, or if such a change can reasonably be anticipated, that it will advise your organization immediately upon identification of that possibility.

Further, the supplier's accounting methodologies should be consistent with generally accepted accounting principles (GAAP), and evaluated by an independent, recognized, certified public accounting firm, to confirm that its financial reporting meets those GAAP requirements.

Financial stability may not always be evident from the balance sheet, or the supplier's profit-and-loss statements, so be cautious. For example, privately-owned firms may elect to reinvest their excess profits in R&D, employee gain-sharing plans, bonuses, or other devices that shift or lower tax burden. Their profit-and-loss statements, in those cases, will often show a moderate profit or no profit at all. At first glance, you would consider such firms to be shaky or too small to support your needs for the long term. Due to the potential risks involved in dealing with such financially challenged suppliers, you might elect to place your business with other, more financially stable firms. But with further due diligence, you may find that a supplier in question consistently generates sufficient cash flow from continuing operations to support both the business and technological growth required to support your dynamic requirements. Don't rely solely on the traditional financial reporting mechanisms. Dig deeper. And under no circumstances should you rely solely on the credit and business rating organizations to provide you with financial data. Their information often leaves many questions unanswered.

### Knowledge of the Industry

The suppliers you elect to deal with, along with their employees and representatives, should be capable of demonstrating that they are knowledgeable of your industry, as demonstrated by their current product lines, services, and customer base. The suppliers and their employees should be active in the industry, as evidenced through their involvement in industry trade groups, technical associations, and industry publications. In short, suppliers can demonstrate their commitment to long-term growth within your industry segment because they have consistently committed their resources to developing capabilities within it.

Suppliers, because of their industry involvement, should be capable of consistently providing your product and service designers with critical insight into current and projected market and industry trends, as well as any anticipated technological developments that could shift market advantage to your competitors. This is not intended to indicate that you should seek confidential or sensitive information about your competitors' specific operations or technologies from these suppliers. It is, rather, the ability to rely on their industry knowledge to provide general guidance relating to movement or shifts in their markets that may impact yours.

*Facilities Management and Maintenance*
As you know from your own operations, unplanned downtime and inefficient equipment can and will lead to reduced productivity, quality, and delivery performance. It is therefore essential that you deal with suppliers that demonstrate an effective and sustained commitment to the maintenance and improvement of both their production facilities and equipment.

Such suppliers, either existing or potential, should be able to demonstrate that their production facilities and equipment are maintained to ensure continuous, uninterrupted delivery of products and services to your organization. Where mandated by regulatory agencies such as the FDA, the suppliers should have records that demonstrate that they follow "good manufacturing practices." They also should be capable of verifying through documented records that their maintenance procedures incorporate both preventive and predictive maintenance practices and methodologies. In addition, they must demonstrate that their equipment and facilities maintenance schedules are routinely followed, and that results of those maintenance activities are used to implement corrective action processes to ensure that their process controls are maintained. Finally, the supplier's equipment and facilities management processes should contain provisions for safety-related monitoring programs to ensure a safe workplace for all employees.

*Distribution Process Control*
Suppliers should be able to demonstrate that their shipping and distribution processes are effectively controlled and managed. Such

control ensures that their customers receive the correct products and the exact quantities of those products consistent with their purchase orders, and that those products are delivered to the designated customer destination on time. Suppliers should also be capable of demonstrating that their packaging, handling, and transportation methods ensure that all products are consistently delivered in good condition, free of in-transit or handling damage. The supplier can demonstrate that they comply with all customer and DOT requirements by furnishing accurate, audited shipping documentation such as freight bills, invoices, and packing slips, marked or coded as designated by the customer. In addition, if you so require, the supplier should be able to demonstrate that it is capable of supporting recognized bar coding and electronic data interchange (EDI) standards.

### Order-entry Process Control

A frequently overlooked activity in the supply-chain management process is a review of the supplier's order-entry processes. I have discovered that it is not uncommon to find suppliers introducing as much as a 20–25% error rate during these processes and consuming as much as 60% of the total quoted lead time. It is important, then, that you consider this up-front administrative function relative to its impact on quality and delivery performance, as well as the total lead time from the placement of your order with the supplier to its delivery to your dock.

The supplier should be capable of demonstrating that its order-entry processes are routinely monitored by management through formal, customer-focused performance metrics to ensure the accurate and timely processing of customer orders. Typical customer-focused order-entry performance metrics target:

- Order-entry cycle time,
- Customer satisfaction,
- On-time delivery performance to promise,
- Order-entry accuracy or error rates,
- Cycle time for resolution of customer complaints,
- Customer inquiry response time, and
- Order pricing accuracy.

Suppliers also must be capable of demonstrating that their order entry metrics are used by their senior management, operations personnel, and marketing/sales management to implement such continuous improvements in the order-entry process that have been proven effective in enabling the supplier to better serve its customers. Evidence of those corrective action results should be documented through formal measurement techniques and confirmed by the supplier's customers.

### Customer and Field Service

The supplier can demonstrate that its customer service and/or field service functions are clearly defined, monitored, and supported by management, as evidenced by direct communication channels to senior management and the use of customer-focused performance metrics to monitor customer service and satisfaction levels. Those performance metrics should further demonstrate that the supplier's customer service and field service functions consistently provide prompt response to customer requests for information or on-sight assistance (measured in hours or less) through direct lines of communication with their customers (mail, fax, EDI, telephone).

The supplier can demonstrate that it is supportive of flexible delivery schedules that are consistent with the needs of your organization, along with lot sizes that are consistent with your inventory policies, without additional costs or up-charges.

### Compliance with Governmental Regulations

Any supplier you deal with *must* be capable of demonstrating that it is in compliance with all applicable federal, state, and local regulatory requirements, including:
- Equal Employment Opportunity Commission (EEOC),
- Occupational Safety and Health Administration (OSHA),
- Environmental Protection Agency (EPA),
- Internal Revenue Service (IRS), and
- Division of Employment Security.

The last thing you want is to be dragged into a class-action suit or have your single-sourced supplier closed by the government unexpectedly for failure to comply with a regulatory requirement.

Each supplier must have documented procedures, policies, and practices in place which ensure environmentally benign operations, employee safety, and compliance with government mandated employee selection and promotion criteria. And, of course, payment of all applicable payroll and federal income taxes is always good practice. The supplier must further be capable of demonstrating that it possesses documentation stating that products provided to your organization are in strict compliance with all applicable regulations, and that any products containing materials or chemicals covered under said regulations are labeled appropriately, pursuant to applicable regulations and laws.

*Labor Relations*
Contented employees produce quality products. Unfortunately, the converse is also true. It is therefore important that existing or potential suppliers be capable of demonstrating that their management promote a stable and productive working environment for their employees, utilizing labor-management techniques that focus on involving all employees in the day-to-day management of their companies. Evidence that the supplier provides training and educational opportunities for employees at all levels to enhance their skills and decision-making capabilities should be readily available and adequately documented. And, of course, the supplier must be able to provide documented evidence of its compliance with all applicable EEOC guidelines.

In addition, the supplier should be capable of demonstrating that safety programs are employed that have been proven to be effective in minimizing lost time injuries and doctors' cases. Further, the supplier should be able to demonstrate through documentation that its labor relations are sound and productive, and as such, pose no threat to delivery schedules or quality levels for the products and services provided to your organization.

*Document Control*
The supplier should provide evidence of a formal document control process which guarantees that all customer specifications, bills of material, engineering changes, drawing revisions, and supporting documentation are current and readily available to appropriate operating personnel. Suppliers' document control processes should be contained in written procedures which are included as part of its employee training programs. The supplier also should have evidence that appropriate management-level personnel routinely confirm the validity of customer documentation in conjunction with its customers, as well as upon receipt of an order for products or services (prior to commencing work on the order).

*Logistics*

With the improvement of surface and air freight services during the past decade, the geographic location of a supplier is no longer a critical issue. However, the supplier's facility must be located near main transportation arteries (truck, air, rail, port) to ensure timely and economical delivery of your products or services. The supplier's ability to deliver products and services should not be contingent on or impaired by bad weather or other logistical deterrents.

The supplier should be able to demonstrate competence in dealing with international customs and duty requirements, should shipments be required to cross international borders. In addition, the supplier should provide flexibility in assigning the FOB point for your shipments to coincide with the requirements of your internal traffic operations.

*Product Reliability and Warranty*

Make sure your supplier is capable of demonstrating that its product reliability has been confirmed through FMEA, DOE, or other applicable analyses in applications similar to that routinely used in your operations or those of your customers. The supplier should be capable of not only providing documented results of those analyses, but be willing to provide copies of them upon request for review by your organization.

The supplier should be able to demonstrate its policy of providing industry-standard (or better) warranty terms for the products or services it provides. Those policies should define how the supplier employs a formal process for handling, managing, and monitoring warranty claims. Documented evidence should be available which confirms that the supplier's warranty claim processing conforms to the supplier's internal performance targets, as well as those of its customers.

### Developing the Audit Assessment Tools

Once you have assessed the minimum requirements for your supply-chain management process, developed a comprehensive description for each, and critically evaluated their relevancy to your operation and customer requirements, you are ready to take the second step in the process: expanding those requirements into a quantifiable supplier assessment tool. Such a tool would be used during an on-site audit to determine the actual compliance levels of a supplier's operations based on those requirements.

The key to any good audit tool is that it be user friendly. The questions covered in the audit process should be focused only on those issues that are relevant to the products or services provided to you, should not be prescriptive in nature, and should be easily measurable to ensure the maximum objectivity possible.

To illustrate, we'll take a look at a few examples of typical audit questions used in world-class supply-chain management programs.

*Manufacturing Process Capabilities and Capacities*
- Can this supplier demonstrate that it conducts regular reviews of its internal process capabilities with the objective of expanding and enhancing those capabilities?
- Is there evidence that the results of this supplier's process capability studies are forwarded to its and our design engineering staff for use in its and our product development activities?
- Is there evidence that this supplier uses recognized problem-solving and decision-making techniques to identify, measure, and resolve internal and external process problems?
- Are those problem-solving and decision-making techniques documented and applied in a timely manner to minimize the impact of the problem?
- Can the supplier demonstrate that those problem-solving and decision-making techniques have produced results that have systematically reduced process variability with a goal of 100% first-pass yield?
- Is there evidence that this supplier employs statistical techniques to continuously monitor process capability against our product specifications?
- Is there evidence that this supplier provides comprehensive procedures to its operators that clearly define the actions to be taken to prevent a process from moving out of control?
- Is there evidence that this supplier employs design of experiments (DOE) or other preventive techniques to design out potential variability in its processes?
- Is there evidence that this supplier obtains process capability studies from its key suppliers prior to placement of orders with them?

- Is there evidence that this supplier utilizes the process capability studies in determining which suppliers to approve for critical component purchases?

*Manufacturing Process Control*
- Is there documentation to confirm that all areas of the supplier's facility are kept clean and free of nonessential inventory, tools, dies, fixtures, etc.?
- Is there evidence that the supplier has proper storage and control procedures for any hazardous materials used in the production process?
- Does the supplier have a written procedure for electrostatic discharge protection when electrical components are used?
  - Can the supplier demonstrate that those procedures are followed by all appropriate personnel involved with the production, handling, storage, and transportation of those products?
- Is there evidence that the supplier utilizes a "pull" versus "push" technique to drive production (i.e.: *kanban* or final assembly order processing)?
- Can the supplier demonstrate that their manufacturing lots are traceable throughout their production processes?
- Can the supplier provide evidence of a written procedure defining the statistical process control methods they employ?
  - Does it define the methods of reporting?
  - Does it outline the frequency and timing of samples?
  - Does it include the maintenance of statistically-based control charts?
- Is there evidence that the supplier has established process controls that are maintained at all critical points within the process as defined by either the supplier's or their customers' process or design engineers?
- Is process control data documented and distributed often enough to provide an early warning of impending process control problems?
- Can the supplier demonstrate that its process control data triggers corrective action when the process is not within control limits?
  - Is the corrective action process clearly defined along with the responsible parties?

- Can the supplier demonstrate that such actions have yielded positive, measurable results for the customer?
- Can the supplier show evidence that process changes are controlled, authorized, documented, dated, and signed?
- Is there evidence that process and product specifications are readily accessible to all appropriate operators?
- Is there evidence that all operators are trained and capable of interpreting customer specifications?
- Is there a documented policy or procedure that defines the conditions under which operators and maintenance personnel have the authority to shut down production operations?
    - Under all out-of-control conditions?
    - Is there evidence that the procedure is understood and acted upon in the presence of an out-of-control or potentially out-of-control situation?
- Does the supplier maintain documented process and inspection records for a period consistent with the requirements of either our industry or any unique requirements of our company?
- Is there evidence that process and inspection records are accessible to all appropriate operators and quality personnel?
- Is there evidence of a written procedure for process audits?
    - Does the procedure specify methods of reporting findings and recommendations?
    - Does it define methods for corrective action and the responsible parties?
- Is there evidence that SPC or inspection charts are regularly maintained on the shop floor and reviewed by manufacturing and quality management?
- Does the supplier have a documented rework procedure that is consistent with that of either its customers or applicable industry standards?
- Does the supplier have a documented setup procedure?
- Is there evidence of activities designed to reduce setup time and costs?
    - Is there evidence that setup reduction activities have yielded measurable results for the customers?
- Is there evidence that production equipment is consistently calibrated on a regular basis by trained personnel?
- Does the supplier keep accurate calibration records and dates?

- Are procedures in place to quarantine nonconforming materials?
- Are procedures in place to confirm the acceptability of all products prior to release to the customer?
- Is an official engineering or spec change system in place to inform customers of changes to the products in advance?
- Is there evidence that the supplier surveys its customers for approval of any product changes that impact form, fit, or function before the changes are implemented?
- Is there a written procedure for communicating and distributing spec and engineering changes?
- Can the supplier demonstrate that obsolete engineering drawings and product specifications are removed from all manufacturing and quality assurance operations immediately upon implementation of a new revision?
- Can the supplier demonstrate that all documents and drawings that are used in the manufacturing process are free of unofficial and handwritten changes?
- Does the supplier have a properly documented and enforced preventive maintenance system?
  - Predictive maintenance system?
  - Are documented results available for review?
- Is there evidence that all tools and fixtures used in production are fully qualified, maintained, and identified?
- Does the supplier utilize a documented tool and fixture location system for all production tooling?
- Can the supplier demonstrate that maximum tool life is identified and monitored through formal systems and that the tool life is communicated to all appropriate operators?

*Quality Systems*
- Does the supplier have in place and follow a formal quality system?
- Is the supplier's quality system completely and clearly documented?
- Is there evidence that the supplier's quality system documents all applicable functions that contribute directly to the:
  - Receipt of materials?
  - Purchase of materials?
  - Design of materials?

- – Production of materials?
- – Handling of materials?
- – Storage of materials?
- – Distribution of materials?
- Is there evidence that the supplier's quality system is based on statistical methodologies that comply with applicable national or international standards such as ANSI, ASTM, UL, CSA, ISO 9000, and QS 9000?
- Is there evidence that the supplier's quality system confirms through quantitative methodologies that our specifications and requirements are visible throughout its production or service operations?
- Does the supplier's quality system clearly designate who is responsible for the control and release of all nonconforming materials?
  - – Does the supplier's quality system document the conditions under which nonconforming materials can be released for further processing?
  - – Can the supplier demonstrate consistent adherence to those policies and procedures?
- Is there evidence that the supplier's quality system dictates that any processed materials identified as defective must be scrapped or reworked under guidelines approved by our organization?
- Is there evidence that the supplier's quality system imposes a requirement for timely corrective action responses to the customer when a defect is encountered?
  - – Can the supplier demonstrate in day-to-day operations adherence to that requirement?
- Is there evidence that the supplier's quality system defines the frequency of calibration for inspection and test equipment?
  - – Is that frequency consistent with recognized national standards for the industry?
  - – Are the calibrations documented?
- Is there evidence that the supplier's quality system defines the selection, control, and maintenance of all inspection and test equipment?
- Is there evidence that the supplier's quality system includes a requirement for internal audits to ensure that all quality ac-

tivities comply with the supplier's quality system and quality policies?

- – Can the supplier demonstrate that its senior management is actively involved in the review of quality system audit results?
- – Where indicated, can the supplier demonstrate that its management is actively involved in the corrective action processes required to bring its activities into compliance with quality system policies?
- Does the supplier document and retain complete historical information on all shipments or services provided for a period of not less than one year?
- Is there evidence that such historical quality information is used as the baseline for continuous improvement activities?
- Is there evidence that such historical information has been used as the basis for corrective action implementations?
  - – Is there evidence that those corrective actions have been successfully implemented yielding measurable positive results?
- Can the supplier demonstrate that its quality system properly controls all applicable customer specification changes to ensure that only current revision levels are produced and shipped?
- Can the supplier demonstrate that its quality system governs raw material specifications for internal use that are consistent with those of your product and/or service specifications?
- Is there evidence that the supplier's quality system dictates an incoming inspection of raw materials and components, and supplier test certifications prior to the acceptance of raw materials from their suppliers?
- Can the supplier demonstrate that their quality system ensures that all required testing procedures and inspection capabilities are available with which to perform all of our designated and required tests or quality confirmations prior to shipment?
  - – Is there evidence that those test methods are confirmed and controlled regularly through the supplier's quality system?
- Does the supplier's quality system define who has authority to stop the shipment of material?
  - – Is there evidence that the conditions under which a stop shipment order is issued are well understood by the appropriate personnel?

*Quality Management*
- Can the supplier demonstrate that the management of their quality system clearly supports our organization's quality requirements as evidenced by:
  - A formal written quality policy and quality manual?
  - Measurable quality objectives?
  - Quantified quality performance metrics?
  - An organizational structure that clearly defines the lines of authority and responsibility relative to the quality systems and processes?
- Can the supplier demonstrate that it promotes operator control versus conventional in-process inspection by independent quality inspectors?
  - Is there evidence that the supplier's operators are adequately trained in both the quality and technical aspects of their position?
- Is there evidence that inspection procedures are both monitored and enforced by quality and operations management to ensure daily compliance to quality system requirements?
- Is there evidence that the supplier's test procedures and methods cover:
  - How and when samples are to be taken?
  - The proper equipment to be used for each test?
  - How to compare the results obtained during testing to established acceptance criteria?
  - The methods to be employed to record test results?
  - To whom results are to be reported?
  - What action to take if results are outside acceptable limits?
  - How the material is to be released for further processing or testing?
- Is there evidence that the supplier's senior management conducts regular quality review meetings?
  - If so, is there evidence that those quality reviews concentrate on issues such as customer satisfaction?
- Does the supplier have a long-term quality system improvement plan that is documented as part of its strategic plans?
- Can the supplier demonstrate that it is striving to achieve best-in-class status as part of its quality improvement objectives?
- Is there evidence that the supplier's quality improvement plan is incorporated into the individual departmental or functional

objectives for each part of the organizations, with corresponding performance metrics established to monitor progress?

- Is there evidence that the quality improvement plan is effectively communicated to and understood by all levels within the organization?
- Is there evidence that the supplier's quality policy is a "living document" in that the policy is constantly updated and enhanced to reflect changes in customer requirements, process improvements, etc.?
- Can the supplier provide evidence of ongoing quality training for all levels and functions within the organization?
- Is quality training documented in the personnel records of each of the supplier's workers?
- Has the supplier's quality system been certified by a qualified, independent third party or another customer?
  - Can the supplier demonstrate that the criteria used for said certification are compatible with those of your organization?
- Is there evidence that the supplier employs a formal follow-up system to monitor customer complaints and implement corrective actions relative to those complaints?
- Can the supplier demonstrate that its senior management regularly monitors such critical customer performance metrics as on-time delivery, outgoing quality acceptance levels, and count accuracy?
- Can the supplier demonstrate that its senior management takes corrective action when key customer performance metrics are not met?
- Is there evidence that the supplier's operating and quality personnel utilize statistical problem-solving methods to resolve performance problems in both administrative and operating areas?
  - Can the supplier show tangible results?
- Can the supplier demonstrate that it utilizes statistical process control techniques to ensure ongoing process control in both operating and support areas?
  - Is SPC data available for review?
  - Is there evidence that SPC data has been used as the basis for corrective actions?
- Is there evidence that the supplier promotes a partnership-type relationship with its customers and their suppliers?

- Is there evidence that the supplier utilizes quality stamps and formal quality certification documentation to indicate quality acceptance of products prior to shipment?
  - Is there evidence that quality stamps or certifications are effectively controlled?

## Compliance of Raw Materials

- Can the supplier demonstrate that it procures raw materials, finished components, and subcontracted services that meet or exceed our product specifications?
- Is there evidence that the supplier confirms the incoming quality of its purchased materials through incoming inspection methods or through the receipt of its suppliers' material certifications?
- Does the supplier use a documented and metrics-based selection and certification process to qualify its suppliers?
  - Do the performance metrics include:
    - Conformance to quality requirements?
    - On-time delivery performance?
    - Count accuracy?
- Is there evidence that the supplier provides an evaluation of its suppliers' performance against those metrics on a regular basis (monthly or quarterly) to prompt corrective actions on the part of its suppliers?
- Is there evidence that the supplier requires a timely corrective action feedback (within 30 to 60 days) for all defects received from its supply base?
- Is there evidence that the supplier requires written specifications from its suppliers for all supplier-designed materials?
  - Is there evidence that those specifications are verified for both accuracy and the appropriate revision level?
- If the supplier utilizes subcontractors for any in-process operations, is there evidence that the supplier has the necessary controls in place with which to ensure that your quality requirements are met?
- Does the supplier maintain documented records regarding approved suppliers?
  - Are those records available to the supplier's designers and engineers?

- Is there evidence that performance metrics like quality, on-time delivery, and count accuracy are being actively and universally monitored for each supplier of critical commodities?
- Is there evidence that the supplier periodically audits its suppliers using a team of the supplier's employees, or an independent auditing firm retained for that specific purpose?
- Is there evidence that the supplier maintains statistical data on each of its key suppliers' process capabilities?
- Is there evidence that the supplier maintains a written procedure covering the receipt of materials by the supplier?
  - Is there evidence that the procedure is audited regularly by the supplier's quality or internal audit staff?
  - Is there evidence of consistent compliance to those receiving procedures?
- Is there evidence that the supplier's incoming materials are properly segregated from previously accepted materials until the supplier's approval process is completed?
- Can the supplier demonstrate that their incoming materials are properly protected from the environment upon receipt?
- Is there evidence that the supplier employs a procedure that ensures the identification and traceability of incoming lots from receiving throughout its production processes?
- Is there evidence that all of the supplier's raw material testing and confirmation procedures are properly documented?
  - Is there evidence that the raw material test results are monitored at management-level by the supplier?
- Can the supplier demonstrate EDI capabilities that are consistent with industry standard transaction sets?
- Can the supplier demonstrate that they possess bar coding capabilities that are consistent with the format(s) required by our company?

*Inventory Management and Control Systems*
- Is there evidence that the supplier maintains written procedures outlining the storage, release, and handling of raw materials and in-process inventory items?
- Is there evidence that those procedures ensure that only approved materials are released to and used in the production of our products?

- Is there evidence that the supplier maintains and employs an effective inventory control procedure that maximizes its inventory turns while controlling the incidence of inventory obsolescence?
  - Can the supplier demonstrate that those procedures cover issues such as shelf life and first-in-first-out methods to prevent deterioration of materials?
  - Is there evidence that those procedures are effective in maintaining the supplier's current levels of surplus and obsolete inventory to less than 2.50% of the supplier's total inventory in dollars?
- Can the supplier demonstrate that they use either physical inventories or cycle counting to guarantee inventory accuracy?
  - Is there evidence that the supplier's current inventory accuracy meets or exceeds 99% by part count?
- Is there evidence that the supplier utilizes a formal shop order routing method to ensure that in-process materials are produced to predetermined product or customer specifications?
- Is there evidence that the supplier employs automated material planning techniques to forecast and schedule raw materials and customer orders for finished goods?
- Is there evidence that the supplier labels or codes materials to ensure their proper identification in storage and in process?

*Technical and Product Support*
- Can the supplier demonstrate that the scope of their R&D and product development activities is sufficient to support the unique dynamics or technologies of our industry?
- Is there evidence that the supplier's R&D and new product development efforts are directed at the needs of our organization, both now and in the future?
- Can the supplier demonstrate that our product or process specifications are considered during its new product development processes?
- Is there evidence that the supplier utilizes concurrent or simultaneous engineering methodologies during the development of its new products and processes?
- Is there evidence that the supplier's new product development processes are clearly defined and proceduralized?

- Is there evidence that the supplier's engineers incorporate their company's internal manufacturing process capabilities into their product and process design rules and criteria using sound design-for-producibility techniques?
  - Similarly, is there evidence of incorporation of its suppliers' process capabilities?
- Can the supplier demonstrate that its design changes are effectively controlled throughout its organization?
  - Is there evidence that the supplier's process is proceduralized?
  - Is there evidence that the number of design changes initiated by the supplier is monitored and managed effectively?
- Can the supplier demonstrate that their engineering and design staff are active in industry-related professional associations and continuing education?
- Is there evidence that the supplier employs computer-based design tools to support the design of its new products and processes?
  - Is there evidence that those computer-based design tools include computer simulation and modeling, design of experiments, and FMEA analyses?
- Can the supplier demonstrate that it supports its designers and engineers in the pursuit of design patents as a means to encourage innovation?
- Is there evidence that the supplier follows a strategic long-term technology plan to guide the continued development of leading-edge technologies, where applicable?
- Can the supplier demonstrate that it employs an effective system for maintaining and storing customer and supplier documentation?
- Is there evidence that the supplier has a formal mechanism for notifying their customers of documentation problems?
- Is there evidence that the supplier's new product and process development procedures adequately define the metrics it will employ to guarantee compliance with the design rules, cycle times, performance requirements, reliability requirements, and other customer-driven performance metrics?
- Is there evidence that the supplier employs a procedure to control critical-to-function characteristics for new products before

initiating production, including SPC characteristics and frequencies?

- Can the supplier demonstrate that its key suppliers are included as part of its design team under normal circumstances?
  - Is there also evidence that key customers are included as part of the supplier's product and process design team?
- Is there evidence that the supplier provides technical support to our employees relative to field service, installation, and applications-related issues?
- Will the supplier provide installation and users manuals which clearly delineate the use and maintenance of its products in our applications?

*Cost Controls*
- Can the supplier demonstrate that it actively addresses product and process cost-containment targeting:
  - Waste reduction in direct and administrative areas?
  - Supply base-related material and service cost controls?
  - Production cycle time and throughput improvements?
  - Labor and equipment efficiency improvements?
  - Product development cycle-time improvements?
  - Administrative cycle-time improvements?
- Is there evidence that the supplier has established product cost standards against which operational metrics can be applied?
- Can the supplier demonstrate that all appropriate operational metrics are tracked daily?
  - Is there evidence that all significant variances to standard are reviewed and corrective actions implemented?
- Is there evidence that the supplier's product cost standards are compatible with all quality, operational, and product performance metrics?
- Can the supplier demonstrate that overhead and burden costs are accurately applied to the products and services sold to your organization?
- Is there evidence that the supplier's accounting functions interact frequently and supportively with the supplier's operational functions?
- Is there evidence that the supplier tracks its cost of quality, or a similar quality cost metric?

- Is there evidence that appropriate corrective action plans are implemented when dictated by the result of a poor cost-of-quality measurement?
- Is there quantifiable evidence that those implementations have yielded positive results in maintaining or reducing quality costs?
- Is there evidence that the supplier effectively controls overtime or other lead time-associated costs?
- Is there evidence that the supplier effectively controls inventory-related costs that impact raw materials, work-in-process inventories, and finished goods?
- Is there evidence that the supplier's labor costs are effectively controlled?
- Can the supplier demonstrate that its cost controls have been successful in reducing product or service costs during the prior 12 to 24 months?

*Commitment of Management*
- Is there evidence that the supplier's management is committed to continuous quality improvement?
- Is the supplier's organizational structure well defined and documented?
- Can the supplier demonstrate that its senior management supports and promotes new ideas and concepts for continuous improvement?
- Is there evidence that the supplier's management provides training consistent with employees' job functions and level of responsibility?
- Is there evidence that the supplier's senior management is actively involved in the pursuit of quality and process improvements that directly impact customers?
- Can the supplier demonstrate that their management supports and promotes the concept of employee involvement in the resolution of operational problems and the improvement of quality and process deficiencies?
- Is there evidence that the supplier's senior management has identified and implemented provisions for special controls, tools, employee skills, and processes required to guarantee consistent product quality improvements?

- Is there evidence that the supplier's senior management has established formal written objectives and initiatives aimed at reducing the cycle time of all administrative functions such as order entry, new product development, response to customer complaints, and requests for information, purchasing, material planning, and production scheduling?
- Is there evidence that the supplier's senior management reviews customer-focused organizational performance metrics at least weekly on such performance issues as customer satisfaction, on-time delivery, out-going quality levels, shipment count accuracy, quoted and requested order lead times, etc.?
  - Is there evidence that the supplier's senior management initiates corrective actions when results do not meet objectives?
- Is there evidence that these actions have been successful?
- Can the supplier demonstrate that it is doing business with other customer(s) in a partnership-type relationship as defined by our organization?
- Is there evidence that the supplier's senior management is receptive to innovation and improvement suggestions from both its employees and customers?
  - Its suppliers?
- Is there evidence that the supplier's senior management regularly shares operational and customer performance results with its employees?
- Is there evidence that the supplier's senior management uses the results of internal quality system audits to implement corrective actions and quality system enhancements?
- Is there evidence that the supplier routinely notifies its customers of potential nonconformances or late deliveries in advance of the scheduled due date?
- Is there evidence that the supplier's senior managers personally visit customers and suppliers on a regular basis (at best annually) to solicit input on both product and process improvement ideas?
- Is there evidence that the supplier's senior management maintains a dynamic business plan that extends 3 to 5 years into the future?
  - Is there evidence that the business plan is supported by companion product and integrated facility plans?

- Is there evidence that progress against these plans is monitored?
• Can the supplier demonstrate that its senior management promotes employee involvement in professional organizations such as NAPM, ASQC, SME, SAE, APICS, or other industry-sanctioned trade or professional organizations?
• Is there evidence that the supplier's senior management has adopted a company mission statement developed from documented customer requirements using voice-of-the-customer, quality function deployment, or similar techniques to isolate and prioritize actual customer requirements and expectations?
• Can the supplier demonstrate that its senior management has incorporated its mission into all operating procedures and metrics, at all levels of the organization?
  - Do these procedures and metrics include:
    - How control is established in indirect and direct functions?
    - How customer orders are to be controlled and processed?
    - How product or service defects are to be addressed?
    - How employees are to be trained?
    - How products and processes are to be tested?
    - How information is to be processed and controlled?
    - Who is responsible for quality?
      Management's role in ensuring that all customer-driven quality requirements are met?
    - Management's periodic auditing of the quality systems to guarantee their continued suitability and effectiveness?
    - What is expected from each functional area relative to quality and how it will be measured?
    - Detailed instructions on how management's quality policies and procedures are to be implemented?
    - The handling of technical data, specifications, and control parameters?

*Financial Stability*
• Is there evidence that the supplier is financially stable, as defined by generally accepted accounting principles (GAAP)?
  - Is there evidence substantiating that position from an independent certified public accounting firm?

- Is there evidence that the supplier has experienced a consistent trend of continuous growth in sales and/or profitability during the past five years?
- Is there evidence that the supplier's cash flow from continuing operations is sufficient to sustain normal business activities through the effective utilization (versus sale) of its assets?
- Is there evidence of an appreciable change in ownership during the past 5 years or is there evidence that one is anticipated?
- Is there evidence that the supplier is highly leveraged due to its debt structure, a previous LBO, or other contributing factors or declared obligations?
- Is there any indication that, because of the supplier's cash, technological, or market position, it is a likely candidate for takeover?
- Has the supplier demonstrated a history of frequent or unannounced price increases, using increasing costs or recessionary business conditions as the reason?
- Is there evidence that the supplier has reinvested capital back into the business on a regular basis in an amount equal to or greater than 5-10% of its annual sales revenues?

*Knowledge of the Industry*
- Can this supplier demonstrate that its representatives and employees are familiar with your industry?
- Is there evidence that this supplier provides products and/or services to other companies in industry?
- Is there evidence that this supplier is active in industry trade groups, technical associations, or trade publications in our industry?
- Does this supplier provide insight into market and industry trends, as well as industry or competitive technological developments?
- Is there evidence that this supplier has dedicated resources and investment into servicing our markets?
- Is there evidence that this supplier is vulnerable to forces which would interfere with their continued participation in our industry?
- Is there evidence that a significant percentage of this supplier's business (greater than 30%) comes from our industry?

- Is there evidence that this supplier has been awarded patents or royalties for products developed within our industry within the last 5 years?
- Is there evidence that this supplier is working toward best-in-class status, or other recognizable quality standards within our industry?

*Facilities Management and Maintenance*
- Is there evidence that the supplier maintains and monitors generally good housekeeping practices for every operating department?
- Is there evidence that comprehensive housekeeping audits are scheduled and conducted on a regular basis?
- Is there evidence that the supplier employs an active preventive and predictive maintenance program in all applicable operating areas and facilities?
- Can the supplier demonstrate that all preventive and predictive maintenance schedules are adhered to by all areas and by all levels of management and supervision?
- Do written procedures for reporting deviations from the preventive maintenance standards exist and are they followed?
- Is there evidence that the supplier maintains maintenance records for each piece of operating and test equipment?
  - Is there evidence that maintenance records are accessible to all appropriate operators and maintenance personnel?
- Is the responsibility for the administration and management of the supplier's preventive and predictive maintenance programs clearly documented and understood?
- Can the supplier demonstrate that its material storage areas, aisles, production areas, inspection areas, receiving areas, and shipping areas are clean and free of excess materials, tools, fixtures, and packaging dunnage?
- Can the supplier demonstrate that its housekeeping, safety, and hygiene practices comply with industry standards and any applicable regulatory requirements?
- Is there evidence that timely corrective actions were initiated in response to workplace injuries and lost-time accidents?
- Do written safety instructions (safety manuals, warnings, etc.) exist for all production, material handling, receiving, and shipping operations?

- Are written safety instructions in place for special processes or events such as the handling and disposal of hazardous wastes, fires, emergency weather conditions, floods, earthquakes, etc.?

*Distribution Process Control*
- Is there evidence that the supplier maintains accurate and up-to-date packaging and labeling specifications?
- Can the supplier demonstrate that the responsibility for all packaging specifications and designs is clearly defined and understood by all affected employees?
- Do written procedures exist and are they are effectively deployed in the packaging and shipping departments?
- Is there evidence that the supplier's customers are permitted to define package and label specifications?
- Is there evidence that written procedures exist that define the methods for marking containers for shipment in compliance with customer specifications?
- Can the supplier demonstrate that proper identification labels are consistently used for each package or container in a shipment in accordance with the customer's specifications or any applicable U.S. or international regulatory requirement?
- Can the supplier demonstrate that customer routing and traffic instructions are always available within the shipping and packaging departments?
- Can the supplier demonstrate that it possesses and provides bar code technology consistent with the customer's labeling requirements?
- Is there evidence that the responsibility for company logistics is clearly defined and documented?
- Is there evidence that the supplier maintains agreements with domestic and (where applicable) international freight carriers to ensure consistent on-time delivery performance at the most advantageous rates to their customers?
- Is there evidence that the supplier can track inbound and outbound freight by EDI or other electronic techniques?
- Is there evidence that the supplier provides a certificate of compliance to customer specifications and test requirements with all shipments to its customers?

- Is there evidence that all customer requests are honored for special marking of part numbers, work order numbers, PO numbers, and material codes on outgoing containers for ease in identification?
- Is there evidence that the supplier maintains control over lots required to be kept intact throughout the packaging and shipping processes?
- Is there evidence that all nonconforming materials found during the packaging and shipping processes are properly identified and quarantined from known good material to prevent inadvertent shipment to the customer?
- Is there evidence that procedures exist that restrict the storage and release of materials for shipping only to authorized personnel?
- Can the supplier demonstrate that it maintains a shelf-life program for distributed products where applicable?
- Is there evidence that the supplier maintains written agreements with transport companies and forwarding agents which include procedures to return defective or faulty materials, as well as a process for handling materials that are not received?
- Is there evidence that the supplier has a documented returned-goods policy and procedure?
- Can the supplier demonstrate that it supports a ship-direct request from its customers and that these services are adequately documented and provided?
- Can the supplier demonstrate adequate ability to track shipments from point of origin to destination as part of its normal operations and services?
  - Is there evidence that these practices and procedures are supported by on-line, real-time computer systems?
- Is the supplier able to provide evidence that material ordered for same-day shipment (UPS, Federal Express, etc.) is, by procedure and practice, documented to reflect the actual date shipped?
- Is there evidence that the supplier will make special drop shipments for "truckload" and "less-than-truckload" quantities for both single and multiple locations?

*Order-entry Process Control*
- Is there evidence that supplier maintains written procedures that govern the order-entry process?

- Is there evidence that those procedures are based on customer-designated performance criteria?
- Is there evidence that supplier's senior managers routinely monitor the performance of their order-entry process to make certain it complies with customer expectations of accurate and timely processing of orders?
- Can the supplier demonstrate that its order-entry accuracy rate is continuously measured?
  - Can the supplier demonstrate that its order-entry accuracy rate consistently meets or exceeds an accuracy level that ensures conformance to customer expectations?
- Is there evidence that the supplier measures its order-entry cycle time as one of its primary performance metrics?
  - Can the supplier demonstrate that its targeted order-entry cycle time is consistent with its customers' expectations?
- Is there evidence that the supplier monitors its customer satisfaction levels on a routine basis? If so, does this monitoring include:
  - On-time delivery performance?
  - Outgoing product quality levels?
  - The cycle time for customer complaint resolutions?
  - The ratio of satisfactory complaint resolutions to the total number of complaints received?
  - The ratio of invoicing or pricing errors to the total number of customer orders received?
- Can the supplier demonstrate that the monitoring of its customer satisfaction and order-entry metrics has led to the successful implementation of measurable process performance improvements?
- Is there evidence that the supplier can support the receipt of customer orders via industry-standard EDI transaction sets?

## Customer and Field Service
- Is the supplier's customer service function clearly documented within its existing organizational structure?
- Is there evidence of a formal feedback system to ensure that all customer requests and problems are expeditiously directed to the appropriate department for resolution?
  - Is there evidence that a companion feedback system is in place to ensure that a response from the appropriate inter-

nal department is, in turn, delivered to the customer on a timely basis?

- – Is there evidence that the supplier uses performance targets to govern the customer response time?
- – Is there evidence that performance targets are used to govern the accuracy of those responses?
- Is there evidence that the supplier's senior management is actively involved with the customer service function?
- Is there evidence that the supplier maintains a system that reports both positive and negative trends in customer satisfaction directly to senior management on a timely basis?
- Can the supplier demonstrate that it maintains an emphasis on exemplary levels of customer service for *all* customers versus merely large or key customers?
- Can the supplier produce evidence that its company compares favorably to its direct competition relative to customer satisfaction indices?
- Is there evidence that the supplier's customer and field service processes have yielded consistent improvements in customer satisfaction throughout the last 3 to 5 years?
- Is there evidence that the supplier has benchmarked its customer satisfaction objectives and targets against recognized best-in-class organizations within the industries that make up its customer base?
- Is there evidence that the supplier's senior managers take decisive action to correct performance problems when their company's customer satisfaction levels fall short of targeted or expected goals?
- Is there evidence that the supplier's senior managers personally meet with key customers on a routine basis to discuss customer satisfaction levels and requirements for improvements?
- Is there evidence that the supplier follows a formal, documented procedure for handling customer complaints?
  - – If so, is the procedure consistently effective?
  - – Is the procedure consistently followed by all customer service representatives?
  - – Does the procedure empower the supplier's customer service representatives to resolve routine customer complaints without seeking management approval?

- Does the supplier have a documented procedure to advise customers of potential delivery, quality, or service problems *in advance* of the scheduled ship date?
- Does the supplier use a documented procedure to advise customers of design, manufacturing, or process changes *in advance* of implementing those changes?
- Can the supplier demonstrate an operating policy of providing timely field support when needed to resolve operational or installation problems?
- Is there evidence that the supplier is actively pursuing lead-time reduction activities?
  - Can the supplier demonstrate that lead-time reduction activities have yielded measurable results for its customers?
- Is there evidence that the supplier's management and staff are committed to and follow ethical business practices?

*Compliance with Governmental Regulations*
- Can the supplier demonstrate that it is in full compliance with all federal, state, and local regulatory requirements applicable to its industry segment?
- Can the supplier demonstrate that it maintains a formal, documented procedure governing environmentally responsible and community supportive (proactive) operations?
  - Is there evidence that the supplier's products and operations are environmentally benign?
- Can the supplier demonstrate that all of its products are in compliance with applicable regulations, and that any of its products that contain hazardous materials or chemicals falling under restrictions are labeled appropriately as mandated by regulation?
- Has the supplier been cited within the last 3 years, or is litigation currently pending, for noncompliance with federal, state, or local environmental, safety, or employment regulations?
- Is there evidence that the supplier maintains documented safety and health awareness programs as an integral part of its normal business operations?
  - Is there evidence that safety and health awareness programs are audited annually to confirm their effectiveness?

- Is there evidence that all of the supplier's personnel have been adequately trained in applicable safety, health, and other governmentally-regulated business and operating requirements?
- Does the supplier maintain a documented substance-abuse program that is consistent with federal and state laws?
- Are emergency procedures documented and in place to respond to crisis situations?
- Is there evidence that the supplier provides MSDS sheets and other safety-related informational documents to its employees?
- Is there evidence that the supplier maintains and promotes an active recycling program in conjunction with its suppliers and customers?

*Labor Relations*
- Is there evidence that the supplier supports and promotes employee involvement activities such as work teams or quality improvement teams?
- Is there evidence that the supplier provides cross-training, job training, and educational opportunities for all employees, at every level within its organization?
  - Does the supplier maintain training records on each employee detailing the courses and educational opportunities afforded each employee?
- Is there evidence that the supplier is an equal opportunity employer?
- Is there evidence that the supplier complies with the Uniform Guidelines on Employee Selection?
- Is there evidence that the supplier has experienced a labor-related work stoppage within the last 5 to 6 years?
- Can the supplier demonstrate that all of its labor disputes have been resolved without outside intervention, arbitration, or mediation?
- Is there evidence that the supplier supports and promotes an active employee suggestion program?
  - Is there evidence that a formal policy or procedure exists governing the employee suggestion program that ensures that all suggestions receive objective consideration and recognition?
  - Is there evidence that the procedure ensures that employees submitting suggestions will receive feedback relative to

their suggestion within a reasonable time, whether or not their suggestion was accepted?

- Is there evidence that employees at all levels are actively involved in the supplier's day-to-day decision-making activities?
- Is there evidence that the supplier's senior management keeps employees apprised on a regular basis of the financial and market conditions impacting the company?
- Can the supplier demonstrate that it is equal to or better than industry averages relative to employee:
  - Turnover rate?
  - Absenteeism?
  - Productivity?
  - Advancement?
- Is there evidence that the supplier is involved in either pending or active litigation relative to labor disputes, sexual harassment, discrimination, or unfair labor practices?
- Is there evidence that the average length of service of the supplier's employees is reasonably consistent with other companies within the same industry, given the age of the supplier's operations?

*Document Control*
- Is there evidence that the supplier maintains a document control procedure which guarantees that the specifications, bills of material, engineering change notices, drawing revision levels of customer documentation, and its own, are current and readily available to all appropriate operating personnel?
  - Can the supplier demonstrate that all appropriate employees have been trained on the application and scope of the document control procedure?
  - Is there evidence that those procedures are routinely audited to ensure accuracy?
- Is there evidence that the supplier reviews and confirms all applicable documentation (customer and internal) prior to the processing of each customer order?
- Can the supplier demonstrate that it regularly monitors the accuracy of all internal and customer design and process documentation to ensure a high level of accuracy?
- Is there evidence of a written procedure that defines and monitors the supplier's retention period for key documents?

- Is there evidence that the supplier maintains an effective storage and retrieval process and procedure for all key documents?
- Is there evidence that the supplier maintains an off-site or secured storage location for key documents or backup copies?
- Is there evidence that the supplier restricts or otherwise controls access to key documents?
- Is there evidence that the supplier complies with applicable ISO 9000 document control requirements?

*Product Reliability and Warranty*
- Can the supplier demonstrate that its product reliability and life-cycle costs have been confirmed through FMEA, design-of-experiments, or other statistically-based methodologies?
- Can the supplier demonstrate that it routinely uses design-for-producibility, -assembly, and -maintainability methodologies?
- Can the supplier demonstrate that it maintains documented design rules encompassing those methodologies that have been adopted in day-to-day operations?
- Is there evidence that the supplier routinely tracks product life cycles and life-cycle costs?
- Can the supplier demonstrate that its products routinely perform to advertised specifications, performance levels, and life-cycle costs under applications and environments that are typical to your products?
- Is there evidence that the supplier's product warranties cover all applications, conditions, and environments normally experienced by your products and systems?
- Is there evidence that the supplier's warranty periods, terms, and conditions are comparable to normal industry standards?
- Is there evidence that the supplier maintains a formal policy and procedure for handling warranty claims?
  - Can the supplier provide documented evidence of its adherence to those policies and procedures under normal situations?

## The Scoring Mechanism

Like the self-assessment, the process audit and focused audit tools must contain a user-friendly scoring mechanism with which your auditors can quantitatively evaluate all suppliers within a commodity on the same basis. This will help ensure objectivity.

The 0-5 scale is one of the most commonly-used methods. Other scoring mechanisms of 1-3 or 1-10 are less common, but occasionally used. The key with any scale you select, is to guarantee that each auditor can effectively use the scoring mechanism in much the same way as another auditor under the same or similar circumstances.

To ensure objectivity, it is recommended that the auditors receive hands-on training in the use of the process audit tools and the associated scoring mechanism under the guidance of a seasoned lead auditor. Positive critique will quickly teach your auditors how to effectively assess each topic and assign a supportable score based on their assessment of the supplier's conformance to the scoring criteria.

During an actual process or focused audit, it will be necessary to total the score for each section, then to total the scores for all sections into a final audit score. But in most cases, the level of relative importance of one section versus another must be considered. For example, your organization might place a higher level of importance on a supplier's quality systems than on their handling of warranty claims. In such a case, the quality systems section of the process audit tools would carry a higher weighting than the section addressing warranty processing. In much the same way, you may elect to place a higher level of importance on a supplier's inventory management processes than on its handling of invoices. As before, you would assign a higher weighting to its inventory management processes than you would to its accounts-payable processing. This weighting process holds true throughout the process audit tool.

But be cautious. Do not make the scoring mechanism too cumbersome. Complexity often adds a degree of error, or subjectivity, to the scoring process. Be sure each auditor completely understands and is comfortable with the scoring mechanism before being assigned to an audit. Otherwise, he or she will likely spend more time juggling numbers than assessing the supplier's capabilities and controls.

Figure 8-22 illustrates a typical scoring mechanism used by a manufacturer of electronic components. Notice that the scoring mechanism uses a percentage of assigned points against the total points possible for that section, then multiplies that percentage by the weighting factor. By so doing, the total always equates to a familiar number. In this case, a percentage. The reason why this is so important is that all of us are comfortable with numbers we are familiar with and that we easily understand. We learned to use percentages in school and in business, and because of that, we are comfortable with the "grading" of someone on a "100" scale. We are obviously more comfortable as auditors with such a scale than, for instance, trying to decide if a supplier who scored 467 points did a good or a bad job.

### The Supply-chain Management Core Team

Contrary to what might be intuitive, the supply-chain management process is *not* the sole jurisdiction of either the Purchasing or Quality departments. Based on the scope your organization has established, all functions impacted by or who impact the raw material selection and use processes should be involved to one degree or another in this process. And frankly, that puts the manufacturing engineer right in the middle of things. Manufacturing engineers select and specify process requirements, establish material costs, and frequently source subcontracted processes, so their input and involvement in the supply-chain management process is vital.

In addition, any other functions with direct influence in the selection or use of the organization's raw materials also should be considered as either a part of, or resource for, the core team. Remember our discussion on the power of cross-functionality? The same principles apply here as well. For example, if your objective is a focused assessment of a supplier's manufacturing process control systems, your core team might be comprised of a purchasing agent, a manufacturing engineer, a supplier quality engineer, a quality assurance inspector, a production supervisor, an SPC expert, and a quality systems engineer. On the other hand, if your assessment was targeted at the cost management processes of a particular supplier, your core team might be comprised of a purchasing agent, a manufacturing engineer, a cost accountant, and a financial analyst—all of whom are directly involved in the material costing side of your business.

The role of your core team is three-fold. First, to develop and calibrate the tools with which to conduct the process and focused audits. Second, to train all other auditors on the use of those tools and the associated scoring mechanisms. And third, to lead or take

| Electronics Are Us, Inc. | | | | | |
|---|---|---|---|---|---|
| Process Audit Scoring Summary | | | | | |
| | | | | | |
| Process Audit Section Scoring | Points Awarded | Points Available | Percent | Weight Factor | Score |
| | (A) | (B) | (A÷B=C) | (D) | (C×D) |
| Quality management | | | | 0.12 | |
| Quality systems | | | | 0.10 | |
| Manufacturing process control | | | | 0.08 | |
| Mfg process capability and capacities | | | | 0.08 | |
| Compliance of raw materials | | | | 0.09 | |
| Inventory control systems | | | | 0.09 | |
| Commitment of management | | | | 0.05 | |
| Technical and product support | | | | 0.03 | |
| Cost controls | | | | 0.04 | |
| Financial stability | | | | 0.05 | |
| Knowledge of the industry | | | | 0.03 | |
| Facilities management and maintenance | | | | 0.04 | |
| Order entry process control | | | | 0.03 | |
| Distribution process controls | | | | 0.04 | |
| Customer and field service | | | | 0.05 | |
| Compliance with regulations | | | | 0.03 | |
| Labor relations | | | | 0.02 | |
| Document control | | | | 0.01 | |
| Logistics | | | | 0.01 | |
| Product warranty and reliability | | | | 0.01 | |
| | | | | 1.00 | |
| Total score | | | | | |

*Figure 8-22. A weighted scoring mechanism provides a meaningful number with which to gage a process audit.*

part in the actual audits. We have discussed in detail the first two; now let's take a look at the audit process itself.

### Preparation for and Planning of the Process Audit

As you now prepare to employ your supply-chain management tools in a full-scale process audit, remember that the audit is intended to do two things:

1. Provide a mechanism to quantitatively assess a supplier's performance so that corrective actions can be developed with which to prevent nonconformances in the future, and

2. Obtain objective evidence as to the supplier's effectiveness in maintaining desired levels of performance, and improving upon that performance level in the future.

The scope of the process audit, as we have discussed, is significantly broader than the focused audit. As such, by addressing the process audit process from this point forward, we can effectively cover both.

The scope of the process audit is simply to verify the existence of the necessary elements within the supplier's quality and business systems that ensure the systems' ability to consistently achieve all established customer performance requirements. The audit should include all direct, indirect, and administrative functions that contribute to or influence those quality and business systems. Therefore, the scope and depth of each process audit will hinge on the specific informational needs of the audit team, the relative complexity of the supplier's operations and processes, and the supplier's product mix and the associated volumes of each product in that mix.

The frequency of the process, and subsequent focused audits, will be contingent on:

- The supplier's compliance with established performance requirements and any corrective actions that may have been identified as a result of a recognized nonconformance or as a result of a prior process audit, or
- Any changes in the management or ownership of the supplier's organization, and any significant changes in the technologies employed by the supplier, the supplier's quality systems or methods, the supplier's manufacturing methods, or the supplier's equipment or facilities.

### Selecting the Audit Team

The audit team will ultimately be responsible for all phases of the process audit, with the vested authority to make final decisions regarding the conduct of the audit and any assessment observations and recommendations. The makeup of the audit team, therefore, should be established with care. The audit team should consist of four to six members (for a full process audit, one or two for a focused audit) whose collective skills encompass experience in all areas that will be touched upon during the audit. In addition, it is good practice to include a member with specialized skills or training not commonly found in most auditors; these skills vary, depending on the nature of the areas to be audited. The secret to a successful audit is to always be completely prepared.

Prior to conducting the process audit, the members of the audit team should be thoroughly trained on the use, application, and scoring of the audit tools. In addition, they should be totally conversant in the fundamentals of the audit process. And finally, it is always a good practice for the audit team to learn each of the supplier's processes that will be reviewed during the audit. A good idea is to ask the supplier's representative to give a slide or video presentation of the areas which the audit will cover (prior to the scheduled audit) and explain to the audit team how each process functions and the associated control mechanisms of each.

### Roles and Responsibilities

Each process and focused audit should be conducted under the guidance of a lead auditor whose responsibilities include:

- Recommending audit team members based upon the needed skills and experience required for the particular audit,
- Preparing the audit plan and assigning team members to specific areas of the process or focused audit,
- Establishing and confirming the schedule with the supplier,
- Identifying and arranging for any pre-training for the individual audit team members,
- Overseeing the audit and addressing any open issues that arise during the audit, and
- Preparing and submitting the final audit report, including any requests for corrective action responses.

Assigning specific tasks to the individual team members for the audit is not always as simple as, for example, assigning the manufacturing engineer to review the supplier's manufacturing operations or the purchasing agent to review the supplier's procurement processes. Often, it is more prudent to assign two team members to one area of the supplier's operations. In this way, a second perspective is added which often proves valuable when reviewing the team's findings after the audit is completed. Also, if the team members assigned to a particular area include the functional expert and the internal customer of that process, then a better understanding of how that particular process meets its internal customer's needs will result.

Once the audit assignments have been made, the lead auditor should review the specific documentation requirements for that area with the auditors. Emphasis should be placed upon what *results* are expected to be documented for each process, as well as what supporting documentation will be required for each audit question or section. Further, the lead auditor should discuss the handling of confidential or sensitive information and data, including what should and what should *not* be removed from the supplier's premises. Always respect the supplier's need for confidentiality and secrecy. After all, they have opened their doors to you in a gesture of good faith. Show them that same respect by keeping their trade secrets confidential.

Another hint for the lead auditor is to identify in advance what documentation will be needed for review during the audit, then to request as much of that documentation as possible in advance of the scheduled audit. It saves time on site and provides advance insight into what areas should be explored in more depth during the process audit.

One of the most important planning roles of the lead auditor is that of coordinating the audit schedule with the supplier at a time that is mutually convenient to both the supplier and the audit team. The audit, after all, will cut a significant amount of time from the schedules of the supplier's key personnel. It is only common courtesy, therefore, to request a set of dates that dovetails with the supplier's operating constraints and schedules. The lead auditor also should confirm the agenda for the audit; advise the supplier of the areas to be covered during the audit and by whom; prearrange any required meetings; and handle any administrative issues like attire, required safety equipment, and working hours.

One final note: it is recommended that the lead auditor provide the supplier with a copy of the audit questions 2-3 weeks in advance of the scheduled process audit. This will allow the supplier to prepare the necessary documentation for review and arrange the schedules of the people who will be required to take part in the audit. It does not give the supplier an advantage or allow them to stack the deck. If the supplier tries to prepare for you, no harm is done. If your team has done a good and comprehensive job of preparing the audit tools and does a professional job of assessing the supplier's operations and controls, no amount of preparation will conceal a weakness in the supplier's processes. The cardinal rule of auditing is *show me*. Look behind every answer for evidence of compliance. If the evidence is lacking or weak, the answer may not accurately convey the supplier's compliance with the area being discussed. And if any auditor encounters difficulty getting the information or data they need to accurately assess the supplier's process controls and capabilities, and if the situation is not satisfactorily addressed by the supplier's management within a short period of time, then the lead auditor should exercise his or her prerogative and call a halt to the audit. After all, the schedule was mutually agreed to, the areas covered by the process audit discussed in advance, the timing and resource needs identified several weeks in advance, and the actual questions to be covered provided in advance. If that is not sufficient for the supplier to prepare for your audit team, then there is more wrong than the supplier wants to admit. Pull the audit team and look for another supplier. That one is hiding from its problems.

Upon completion of the audit, the lead auditor should consolidate the documentation and recommendations of the team members, then compile a formal report (like that illustrated in Figure 8-23). The report should outline the results from each area of the process audit, the associated operational or process weaknesses identified within each area, and the requested corrective actions to address each nonconformance. Corrective action requests should not be prescriptive, but rather should be structured to allow the supplier to develop its own methods to address each weakness and nonconformance. As the customer, your interest is in the actual results achieved, not how they were accomplished.

### Audit and Assessment Hints

Professional auditors have a number of "triggers" they use to identify potential problem areas during a process audit. Once identified, they pursue that particular area in depth until they are comfortable that they have accurately and completely isolated all control and process problems. A few of their "tricks of the trade" are included here for your use during a process or focused audit.

### Rework and Rejections

The discovery of an unusual amount of rework or rejected materials is usually a good indication of process capability and control problems. When searching for the source of those problems, look

**AGI Manufacturing, Inc.**
**Supplier Audit Report Sheet**

Supplier _____     Auditors _____
Location _____              _____
         _____              _____
         _____              _____

Telephone _____     FAX _____
Date _____

Items covered by audit            Description
         _____              _____
         _____              _____

| Category | Possible Points | Points Scored | Percent |
|---|---|---|---|
| 1. Quality Systems | | | |
| 2. Quality Management | | | |
| 3. Manufacturing Process Control | | | |
| 4. Mfg Process Capability and Capacity | | | |
| 5. Compliance of Raw Materials | | | |
| 6. Inventory Control Systems | | | |
| 7. Commitment of Management | | | |
| 8. Technical and Product Support | | | |
| 9. Cost Controls | | | |
| 10. Financial Stability | | | |
| 11. Knowledge of the Industry | | | |
| 12. Facilities and Maintenance | | | |
| 13. Order Entry Process Control | | | |
| 14. Distribution Process Control | | | |
| 15. Customer and Field Service | | | |
| 16. Compliance with Regulations | | | |
| 17. Labor Relations | | | |
| 18. Document Control | | | |
| 19. Logistics | | | |
| 20. Product Reliability and Warranty | | | |
| Raw Totals | | | |
| Total Weighted Score | | | |

Corrective Actions Required                                    Date Due
_____     _____
_____     _____

For additional information, refer to attached Survey Data Sheets for each respective survey section. (Includes any documentation used by the audit team to arrive at their conclusions.)

Recommendations
_____
_____

*Figure 8-23. The official record of the process audit should be consolidated in a formal report encapsulating the data compiled.*

for control deficiencies in both work in process and in the raw materials being received. Hard data sources include internal reject reports, receiving reports, complex or illogical material flows, and multiple handling of materials or documents. Look for the presence of a Material Review Board (MRB)—a formal methodology for processing and dispositioning nonconforming materials. This is a true nonvalue-added activity.

One other secret. Look at the aging factor of the reject tickets as an indication of how effectively and how quickly problems are resolved. Most Quality departments keep track of their open reject tickets. Take a random sample and look for the average aging. It could be an eye-opener.

### Bottlenecks

Look for bottlenecks in material or information flows; they fatten process cycle times and costs. The physical collection of materials by work centers or piles of documents on desks are common clues to problems in this area. Other hard data sources include backlog and material aging reports. Look for batch processing of information versus real-time processing. And look for the number of approvals required to process standard documents like purchase orders and engineering change notices.

### Performance Variations

Look for wide gaps between the expected performance levels within normal direct operations and their established standards. Compare standard industry lead times and processing times to what you observe to be occurring within the supplier's operations. Scrutinize production and material variance reports. Look at the overtime being worked. Look for idle personnel. Look for frustrations or overly-burdened personnel. Observe the pace within both direct and administrative areas.

Variances are indicators of potential delays in deliveries, extended lead times, and excessive cost buildups.

### Documentation Errors

Errors and insufficiencies in documentation are often the hidden causes of problems observed in the supplier's production processes and their supply base. Incorrect bills of materials or product specifications, as an example, often lead to material shortages, inventory buildups, surplus and obsolete inventory problems, incorrect product costing, process routing errors, and continuous quality problems.

Look for bill accuracy metrics, shortage reporting, inventory depletion reports, and location errors as signs of pending or actual problems. And don't overlook the order-entry systems. There have been cases where error rates as high as 40% or more have existed in the front-end processes, driving additional inaccuracy and errors into the planning and scheduling activities. In short, if the supplier's processes start with bad information, it is a sure bet they will end the same way.

### Customer Complaints

A short walk through the supplier's Customer Service department often yields a wealth of valuable information about how the supplier actually performs against customer expectations. Be alert for large or extremely busy departments. Typically the larger the department, the more problems exist. Browse through the customer complaint files for signs of numerous common problems and for associated corrective action results. Are the problems actually being resolved, or are quick-fixes being used to appease the customers?

Look at the morale. Look at the pace. Are the employees frustrated or frenzied, or are they content with the supplier's response to customer issues? Look for customer satisfaction metrics. Look for warranty and returns processing costs. They are all indicators of how the supplier's products and services are being received by their customers.

### Inventory Problems

Inventory problems typically are the result of one or more front-end processing problems:

- Poor bill-of-material accuracy,
- Excessive engineering change activity,
- Poor sales and operations planning and coordination,
- Inaccurate process routings,
- Inadequate inventory planning or scheduling systems,
- Poor supplier performance,
- Inaccurate or incomplete customer order processing.

Any or all of these factors can lead to inventory shortages or buildups, with subsequent delivery and cost impacts. Either way, it spells trouble for you as the customer.

*Personnel Problems*

The truth is simple: contented workers make higher quality products. Unhappy or frustrated workers just don't care. Look for the signs: a history of work stoppages or slowdowns, constant grievance problems resulting in arbitration, high absenteeism or turnover rates, or higher than normal safety and health costs.

As you walk through the plant and office areas, observe the people who are unaware of your presence. How are they performing their individual functions? What is their attitude?

There are many, many more secrets you and your team will learn to use as you become accustomed to the audit process. Exchange ideas among the team members. Learn from one another. But never forget: your objective is to assist your suppliers in isolating and resolving their problems to ensure that they consistently perform to your expectations. Your role is neither to dwell on their process weaknesses, nor to overly burden them with insignificant issues that do not impact their ability to perform to expected standards.

In summary, if you control the quality and timeliness of the raw materials entering your manufacturing processes, you will go a long way toward reducing your manufacturing cycle time, increasing your throughput, reducing your operating expenses, reducing your inventory levels, increasing your outgoing quality levels, and thus improving your customer service. And that's what the manufacturing engineer is paid to do.

# 9

# Summary

Plato has said, "People change for one of two reasons, hope or fear." This book speaks of change in numerous areas impacting the manufacturing engineer. These changes will come, with or without our consent. It is our responsibility to be ready, to make the necessary adjustments to our thinking and our skills so that change becomes our ally, not our enemy.

We must lead our organizations into the 21st century with an aggressive, proactive approach that will guarantee our organization's competitiveness in a fiercely competitive and unpredictable global market. Our role in that change process is vastly expanded and eminently more critical than it has ever been in the past.

It's a new world out there, with many more opportunities for manufacturing engineers to explore; many more ways to expand our skills, knowledge, and capabilities. Be ready for the opportunities the 21st century has to offer. Take the lead. Take the challenge. *Be successful.*

# Index